William Saunders

Insects Injurious to Fruits

William Saunders

Insects Injurious to Fruits

ISBN/EAN: 9783743320734

Manufactured in Europe, USA, Canada, Australia, Japa

Cover: Foto ©berggeist007 / pixelio.de

Manufactured and distributed by brebook publishing software
(www.brebook.com)

William Saunders

Insects Injurious to Fruits

INSECTS

INJURIOUS TO FRUITS.

BY

WILLIAM SAUNDERS, F.R.S.C., F.L.S., F.C.S.,

Director of the Experimental Farms of the Dominion of Canada, Fellow of the American
Association for the Advancement of Science, Fellow of the Royal Microscopical
Society of London, England, Fellow of the Entomological Society of London, England, late Editor of the "Canadian Entomologist," Corresponding Member of the American Entomological Society, Philadelphia, of the Buffalo Society of Natural Sciences, the Natural History Society
of Montreal, etc.

ILLUSTRATED WITH FOUR HUNDRED AND FORTY WOOD-CUTS.

SECOND EDITION.

J. B. LIPPINCOTT COMPANY.

LONDON: 10 HENRIETTA STREET, COVENT GARDEN.

1889.

DEDICATION.

To the Fruit-Growers of America this work is respectfully dedicated, with an earnest hope that it may be of practical use to them in the warfare with destructive insects in which they are constantly engaged.

W. SAUNDERS.

PREFACE TO THE SECOND EDITION.

In the preparation of the second edition of this work the author has endeavored to make such corrections, and to embody such additional facts regarding the life history and habits of the insects referred to, and the remedies suggested therefor, as will bring it into accord with the present knowledge of entomologists on these subjects. In this he has been aided by kind suggestions from many friends. Acknowledgments are especially due to C. V. Riley and L. O. Howard of Washington, A. R. Grote of Bremen, Germany, J. A. Lintner of Albany, N. Y., C. H. Fernald of Amherst, Mass., Miss Mary Murtfeldt of Kirkwood, Mo., J. H. Comstock of Ithaca, N. Y., and E. T. Cresson of Philadelphia. The corrections and additions have been embodied in the work without interfering much with its general arrangement.

WILLIAM SAUNDERS.

OTTAWA, Ontario, Canada.

PREFACE TO THE FIRST EDITION.

THE cultivation of fruit in America has of late years become of so much commercial importance, as well as domestic interest, that no apology is necessary for offering to the fruit-growing community a work of which they must have long felt the need.

The amateur who plants a city lot, and the farmer who devotes a portion of his land to the cultivation of those fruits which furnish from month to month pleasant and changeful variety to the table, as well as those who grow fruit to supply the home and foreign markets, are alike interested in making this pursuit a success.

Injurious insects are so universally distributed that there is no part of this continent where fruit-culture can be profitably carried on without some effort being made to subdue them. Among the insect hosts we have friends as well as foes, and it is to the friendly species that nature has assigned the task of keeping in subjection those which are destructive; these, in many instances, do their work most thoroughly, devouring in some cases the eggs, in others the bodies, of their victims. It is not uncommon to find the antipathy to insects carried so far that a war of extermination is waged on all, and thus many of man's most efficient allies are consigned to destruction.

The information necessary to enable the fruit-grower to

5

deal intelligently with this subject has not hitherto been
easily accessible, having been diffused chiefly among a large
number of voluminous State and Departmental reports and
books on scientific entomology, where the practical knowledge
is so much encumbered with scientific and other details as to
make the acquisition of it too laborious a process for those
whose time is so fully occupied during that period when the
information is most needed.

It has been the aim of the author of this work to bring
together all the important facts relating to insects known to
be injurious to fruits in all parts of Canada and the United
States, to add to the information thus obtained the knowledge he has acquired of the habits and life-history of many
of our insect pests by an experience of over twenty years
as a fruit-grower and a student of entomology, and to present the results in as concise and plain a manner as possible,
avoiding all scientific phraseology except such as is necessary
to accuracy.

The arrangement adopted under the several headings, by
which the insect pests which attack the different parts of the
tree or vine under consideration are grouped together, will, it
is hoped, with the aid of the illustrations, greatly facilitate
the determination of any injurious species. When having
before him its history briefly traced and the remedies which
have been found most useful in subduing the insect, the
reader will at once be enabled to decide as to the best methods to be employed.

The author desires to make the fullest acknowledgment to
those of whose work he has availed himself. The writings
of Say, Peck, Harris, Fitch, Clemens, Glover, Walsh, Riley,
Lintner, Comstock, Le Baron, Thomas, French, Packard,

Grote, Leconte, Horn, Hagen, Chambers, Howard, Cook, Uhler, Cresson, Fernald, Kellicott, Willet, Bethune, Pettit, Rogers, Reed, Fletcher, Harrington, and others have been made tributary; and in some instances, where the insect referred to has not been the subject of personal observation, the words of the author drawn from have to some extent been used, modified so as to bring them into harmony with the general aim of this work. To the writings of C. V. Riley, of Washington, the author is especially indebted; his Missouri Reports and subsequent entomological reports in connection with the Department of Agriculture at Washington have been found invaluable.

The material contained in the chapter on orange insects has been derived mainly from the excellent report of J. H. Comstock as Entomologist to the U. S. Department of Agriculture for the year 1880, and from his subsequent writings; from a paper on the parasites which attack scale-insects, by L. O. Howard, in the same report; also from the writings of Townend Glover and C. V. Riley, from a treatise on orange insects, by William H. Ashmead, from a pamphlet on insects injurious to fruit-trees in California, by Matthew Cooke, and from the writings of Dr. S. V. Chapin and others in the first report of the Board of State Agricultural Commissioners of California.

To J. A. Lintner, State Entomologist of New York, the author is under much obligation for his kindly aid in revising the nomenclature. An acknowledgment is also due to the following specialists, who have revised lists submitted to them of the names of insects in their departments: Dr. George H. Horn, E. T. Cresson, A. R. Grote, P. Uhler, J. H. Comstock, and L. O. Howard.

Through the liberality of the Council of the Entomological Society of Ontario, permission was granted to have electrotypes made from any of the cuts in the Society's collection, and from this source a large number of figures have been obtained. Many of these were purchased by the Society from C. V. Riley, and some are the work of Worthington G. Smith, of London, England, and other English and American engravers.

Nos. 21, 22, 31, 93, 102, 104, 116, 137, 141, 142, 145, 169, 199, 201, 205, 206, 291, 292, 305, 321, 332, 347, and 348 were purchased from C. V. Riley.

Nos. 20, 151, 152, 167, and 208 were kindly loaned by A. S. Forbes, of Normal, Illinois.

Through the kind liberality of the Hon. George B. Loring, U. S. Commissioner of Agriculture, permission was granted to obtain electrotypes of the following, which have appeared in the Commissioner's reports : Nos. 13, 15, 32, 35, 42, 96, 108, 114, 115, 126, 181, 195, 248, 270, 286, 287, 288, 377, 393, 394, 400, 403, 404, 406, 407, 408, 409, 410, 411, 412, 413, 414, 416, 418, 419, 420, 421, 422, 423, 424, 426, 428, 429, 431, 432, 433, 434, 435, 436.

Nos. 8, 25, 63, 109, 134, 144, 329, 338, 350, and 401 were purchased from Dr. A. S. Packard.

By kind permission, the following were copied from Townend Glover's excellent plates : Nos. 9, 49, 66, 78, 82, 83, 87, 111, 121, 146, 147, 148, 150, 155, 163, 202, 209, 236, 237, 249, 282, 293, 294, 295, 296, 300, 315, 320, 322, 333, 367, 390, 391, 392, 395, 396, 397, 440.

From Harris's works : Nos. 11, 86, 120, 159, 174, 188.

From the reports of C. V. Riley : Nos. 101, 103, 105, 107, 228, 229, 230, 378, 379.

From the reports of Dr. Asa Fitch : Nos. 36, 37, 98, 99, 301.

From Dr. A. S. Packard's works : Nos. 16, 110, 113, 117, 118, 119, 156, 157, 158, 162, 176, 177, 323, 328, 381, 382, 383, 384, 385, 386, 387, 388.

From B. Walsh's first report No. 143 was copied, No. 55 from one of Cyrus Thomas's reports, No. 187 from a plate published by W. H. Edwards ; Nos. 427 and 430 were copied (reduced in size) from the report of the U. S. Commissioner of Agriculture for 1880, Nos. 438 and 439 from a treatise on insects injurious to fruit-trees in California, by Matthew Cooke, and Nos. 398, 399, 402, 405, 415, 417, 425, and 437 from a treatise on orange insects, by William H. Ashmead.

The remainder have been drawn from nature and engraved for this work chiefly by the following artists, who have also engraved the copies: H. H. Nichol, of Washington ; Worthington G. Smith, of London, England ; H. Faber & Son, and Crosscup & West, of Philadelphia ; and P. J. Edmunds, of London, Ontario.

Throughout this work, where an author's name, following the scientific name of an insect, is enclosed in parentheses, it is an indication that the authority is for the species only, and that the genus has been changed since the insect was described. This is in accordance with the recommendation of the British Association made some years ago, and is now very generally adopted.

WM. SAUNDERS.

London, Ontario, Canada, April 11, 1883.

CONTENTS.

11

INSECTS INJURIOUS TO FRUITS.

INSECTS INJURIOUS TO THE APPLE.

ATTACKING THE ROOTS.

No. 1.—The Apple-root Plant-louse.
Schizoneura lanigera (Hausm.).

THIS insect appears in two forms, one of which attacks the trunk of the apple-tree (see No. 9), the other works under the ground and produces on the roots wart-like swellings and excrescences of all shapes and sizes. These deformities seriously diminish the normal supply of nourishment for the tree, and where very numerous induce gradual decay of the roots, and occasionally result in the death of the tree. Upon close examination the excrescences are found to contain in their crevices very minute pale-yellow lice, often accompanied by larger winged ones. The former have their bodies covered with a bluish-white cottony matter, having the appearance of mould, the filaments of which are five or six times as long as the insects themselves, and are secreted from the upper part of the body, more particularly from the hinder portion of the back. In Fig. 1, *a* represents a knotted root, *b* a wingless louse, and *c* a winged specimen. The insects are both magnified; the short lines at the sides indicate their natural size.

The apple-root plant-louse is believed by some entomologists to be a native insect, while others hold to the opinion that it has been imported from Europe. It is nourished by sucking the juices of the tree, piercing the tender roots with

18

its proboscis.   In the very young lice this instrument, when
at rest and folded under the abdomen, is longer than the
body, but in the more mature specimens it is only about two-
thirds the length of
the body.   While it
usually confines it-
self to the roots of
trees, it is sometimes
found on the suck-
ers that spring up
around them, and
sometimes   also
about the stump of
an   amputated

Fig. 1.

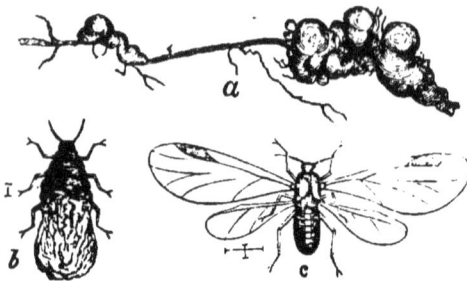

branch, but in every instance it may be recognized by the
bluish-white cottony matter with which its body is covered.
If this cottony covering be forcibly removed, it will be found
that in two or three days the insect will have again produced
sufficient to envelop itself completely.   Occasionally the mature lice crawl up the branches of the trees during the summer, where they also form colonies, and then are known as
the Woolly Aphis of the Apple.   This form of the insect will
be referred to more fully under No. 9.

The appearance of this root-louse is recorded in Downing's
"Horticulturist" as early as 1848, at which time thousands
of young trees were found to be so badly infested that they
had to be destroyed.   Since that period it has been gradually
but widely disseminated, establishing colonies almost everywhere, in the North, South, East, and West.   Where a tree is
sickly from any unknown cause, and no borers can be found
sapping its vitals, the presence of this pest may be suspected.
In such cases the earth should be removed from the roots
about the surface, and these carefully examined, when, if
warty swellings are discovered, no time should be lost in
taking steps to destroy the insidious foe.

Remedies.—The most successful means yet devised for destroying these root-lice is the use of scalding-hot water freely poured around the roots of the trees. If the trees are to remain in the soil, the roots may be laid bare and the water used nearly boiling without injury; but where they have been taken up for the purpose of transplanting, and are to be dipped in the hot water, the temperature should not exceed 150° Fahr.; under these circumstances from 120° to 150° would suffice for the purpose. A mulch placed around the trees for some time previous to treatment has been found useful in bringing the lice to the surface, where they can be more readily reached by the hot water. Drenching the roots with soapsuds has also been recommended, to be followed by a liberal dressing of ashes on the surface.

There are several friendly insects which prey upon the root-louse. A very minute four-winged fly, *Aphelinus mali* (see Fig. 15), is parasitic on it, and the larva of a small beetle belonging to the Lady-bird family, *Scymnus cervicalis*, feeds on it. This friend is difficult to recognize among the lice, from the fact that it is also covered on the back with little tufts of woolly matter secreted from its body; these larvæ are, however, larger than the lice, and much more active, and may be further distinguished by the woolly matter being of an even length, and arranged on the back in transverse rows. The perfect beetle is very small, being but one-twentieth of an inch long, with a dark-brown body and a light-brown thorax. The beetle has been observed preying on lice about the surface of the ground.

A third friendly insect, probably the most efficient check upon the increase of these lice, is known as the Root-louse Syrphus fly, *Pipiza radicum* Riley, which in its larval state feeds upon them. It is then in the form of a footless maggot, which, when full grown, is about a quarter of an inch long (Fig. 2, a), of a dirty yellow color, and usually so covered with dirt and with the woolly matter of the lice it has devoured that it is not easily discerned. The eggs from which these larvæ are produced are laid by the fly (Fig. 2, c) in the spring. The larvæ mature during the summer, and in the fall change to the pupa state, as shown at *b* in the figure, from which the perfect fly emerges the following spring.

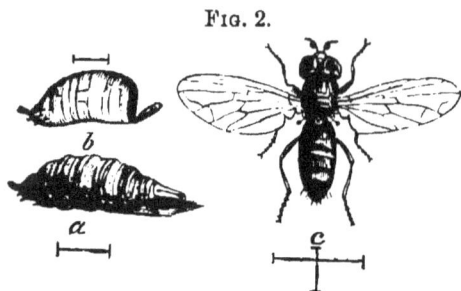

FIG. 2.

The larva, chrysalis, and fly are all magnified in the figure. The fly measures, when its wings are expanded, nearly half an inch across; its body is black, the head hairy with short white hairs, the thorax also similarly hairy and finely punctated; the abdomen finely punctated, and adorned with long white hairs; legs partly reddish, partly black; wings transparent, with black veins.

ATTACKING THE TRUNK.

No. 2.—The Round-headed Apple-tree Borer.

Saperda candida Fabr.

The round-headed apple-tree borer is a native of America, whose existence was unrecorded before 1824, when it was described by Thomas Say. The year following, its destructive character was observed about Albany, N.Y. It is now very widely and generally distributed, and probably it was so at that time, although unnoticed, since it inhabits our native crabs and thorn-bushes, and also the common June-berry, *Amelanchier Canadensis.* While preferring the apple, it also makes its home in the pear, quince, and mountain-ash. In its perfect state it is a very handsome beetle (Fig. 3, *c*), about three-fourths of an inch long, cylindrical in form, of a pale-brown color above, with two broad creamy-white stripes running the whole length of its body; the face and under surface are hoary-white, the antennæ and legs gray. The females are larger than the males, and have shorter antennæ. The beetle makes its appearance during the months of June and July, usually remaining in concealment during the day, and becoming active at dusk.

The eggs are deposited late in June, during July, and most

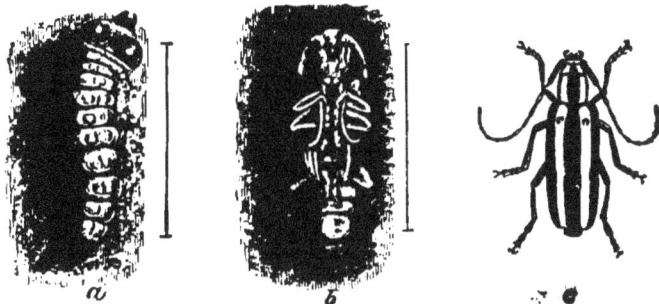

FIG. 3.

of August, one in a place, in an incision made by the female in the bark of the tree near its base. Within two weeks the young larvæ are hatched, and at once commence with their sharp mandibles to gnaw their way to the interior.

It is generally conceded that the larva is three years in reaching maturity. The young ones lie for the first year in the sap-wood and inner bark, excavating flat, shallow cavities, about the size of a silver dollar, which are filled with their sawdust-like castings. The holes by which they enter, being small, are soon filled up, though not until a few grains of castings have fallen from them. Their presence may, however, often be detected in young trees from the bark becoming dark-colored and sometimes dry and dead enough to crack. Through these cracks some of the castings generally protrude, and fall to the ground in a little heap; this takes place especially in the spring of the year, when, with the frequent rains, the heaps become swollen by the absorption of moisture. On the approach of winter the larva descends to the lower part of its burrow, where

2

it doubtless remains inactive until the following spring.
During the next season it attains about half its growth,
still living on the sap-wood, where it does great damage,
and when, as often happens, there are several of these
borers in a single tree, they will sometimes cause its death
by completely girdling it. After another winter's rest, the
larva again becomes active, and towards the end of the
following season, when approaching maturity, it cuts a cylindrical passage upwards, varying in length, into the solid
wood, afterwards extending it outward to the bark, sometimes cutting entirely through the tree, at other times turning back at different angles. The upper part of the cavity
is then filled with a sawdust-like powder, after which the
larva turns round and returns to the part nearest the heart
of the tree, which portion it enlarges by tearing off the
fibres, with which it carefully and securely closes the lower
portion of its gallery, so as to protect it effectually from the
approach of enemies at either end. Having thus perfected
its arrangements, it again turns round so as to have its head
upwards, when it rests from its labors in the interior of the
passage until the following spring, when the mature larva
sheds its skin and discloses the pupa. In this condition it
remains about two or three weeks, when the perfect beetle
escapes. At first its body and wing-cases are soft and flabby,
but in a few days they harden, when the beetle makes its
way through the sawdust-like castings in the upper end of
the passage, and cuts with its powerful jaws a smooth,
round hole through the bark, from which it escapes.

The larva (Fig. 3, *a*) is of a whitish color, with a round
head of a chestnut-brown, polished and horny, and the jaws
black. It has also a yellow horny-looking spot on the first
segment behind the head. It is without feet, but moves
about in its burrows by the alternate contraction and expansion of the segments of its body. When full grown it is
over an inch in length.

The color of the chrysalis (Fig. 3, *b*) is lighter than that of the larva, and it has transverse rows of minute spines on the back, and a few at the extremity of the body.

Remedies.—The young larva, as already stated, may often be detected by the discoloration of the bark. In such instances, if the outer dark-colored surface be scraped with a knife, late in August or early in September, so as to expose the clear white bark beneath, the lurking enemy may be discovered and destroyed. Later they may be detected by their castings, which have been pushed out of the crevices of the bark and have fallen in little heaps on the ground. When first discharged, these look as if they had been forced through the barrels of a minute double-barrelled gun, being arranged closely together in two parallel strings. Those which have burrowed deeper may sometimes be reached by a stout wire thrust into their holes, or by cutting through the bark at the upper end of the chamber, and pouring scalding water into the opening, so that it may soak through the castings and penetrate to the insect.

Among the preventive measures, alkaline washes or solutions are probably the most efficient, since experiments have demonstrated that they are repulsive to the insect, and that the beetle will not lay her eggs on trees protected by such washes. (Soft-soap reduced to the consistence of a thick paint by the addition of a strong solution of washing-soda in water is perhaps as good a formula as can be suggested :) this, if applied to the bark of the tree, especially about the base or collar, and also extended upwards to the crotches, where the main branches have their origin, will cover the whole surface liable to attack, and, if applied during the morning of a warm day, will dry in a few hours, and form a tenacious coating, not easily dissolved by rain. The soap solution should be applied early in June, and a second time during the early part of July.

No. 3.—The Flat-headed Apple-tree Borer.

Chrysobothris femorata (Fabr.).

This borer is also a native of America, and is in its mature state a beetle belonging to the family *Buprestidæ.* It is a very active creature, one which courts the light of day and delights to bask in the hot sunshine, running up and down the bark of a tree with great rapidity, but instantly taking wing if an attempt be made to capture it.   The beetle measures from three-eighths to half an inch or more in length. (See Fig. 4, *d*, where it is shown somewhat enlarged.)   It is of a flattish oblong form and of a shining greenish-black color, each of its wing-cases having three raised lines, the outer two interrupted by two impressed transverse spots of a brassy color, dividing each wing-cover into three nearly equal portions.   The under side of the body and the legs shine like burnished copper; the feet are shining green.

Fig. 4.

This pest is common almost everywhere, affecting alike the frosty regions of the North, the great West, and the sunny South.   It is much more abundant than the two-striped borer, and is a most formidable enemy to apple-culture.   It attacks also the pear, the plum, and sometimes the peach.   In the Southwestern States it begins to appear during the latter part of May, and is found during most of the summer months; in the Northern States and Canada its time of appearance is June and July.   It does not confine its attacks to the base of the tree, but affects the trunk more or less throughout, and sometimes the larger branches.

The eggs, which are yellow and irregularly ribbed, are very small, about one-fiftieth of an inch long, of an ovoidal form, flattened at one end, and are fastened by the female with a glutinous substance, usually under the loose scales or within the cracks and crevices of the bark; sometimes singly, at other times several in a group. The young larva soon hatches, and, having eaten its way through the bark, feeds on the sap-wood within, where, boring broad and flattish channels, a single specimen will sometimes girdle a small tree. As the larva approaches maturity it usually bores into the more solid wood, working upward, and, when about to change to a pupa, cuts a passage back again to the outside, eating nearly but not quite through the bark. Within its retreat it changes to a pupa (Fig. 4, *b*), which is at first white, but gradually approaches in color to that of the future beetle, and in about three weeks the perfect insect emerges, and, having eaten through the thin covering of bark, escapes and roams at large to continue the work of destruction.

The mature larva (Fig. 4, *a*) is a pale-yellow legless grub, with its anterior end enormously enlarged, round, and flattened. At *c* in the figure the under side of the anterior swollen portion of the body is shown. Whether this larva requires one or two seasons to reach maturity has not yet been determined with certainty, but the opinion prevails that its transformations are completed in a single year.

Remedies.—One might reasonably suppose that this larva in its snug retreat would be safe from the attack of outside foes; but it is hunted and devoured by woodpeckers, and also destroyed by insect parasites. A very small fly, a species of Chalcid, destroys many of the larvæ; besides which two larger parasites have been bred from them by Prof. C. V. Riley, one of which, *Bracon charus* Riley, is represented magnified in Fig. 5, the hair-lines at the side showing its natural size. The other species, *Cryptus grallator* Say, is somewhat larger: they both belong to that very useful group of four-winged flies known as Ichneumons.

Although healthy, well-established trees are not exempt from the attacks of this enemy, it is found that sickly trees or trees newly transplanted are more liable to suffer, especially on the southwest side, where the bark is often first injured by exposure to the sun, resulting in what is called

FIG. 5.

sun-scald. All trees should be carefully examined early in the fall, when the young larva, if present, may often be detected by the discoloration of the bark, which sometimes has a flattened and dried appearance, or by a slight exudation of sap, or by the presence of the sawdust-like castings. Whenever such indications are seen, the parts should at once be cut into with a knife and the intruder destroyed. As a preventive measure there is perhaps nothing better than coating the bark of the trunk and larger branches with a mixture of soft-soap and solution of soda, as recommended for the two-striped borer (No. 2).

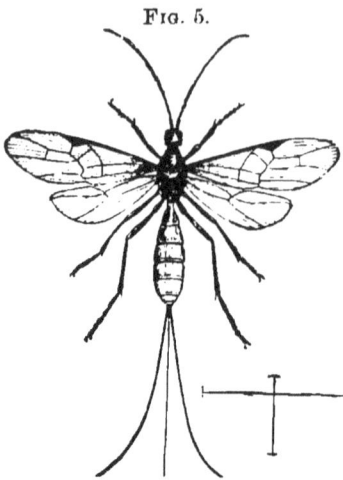

No. 4.—The Long-horned Borer.

Leptostylus aculifer (Say).

Although distributed over a wide area, this is by no means a common insect, and seldom appears in sufficient numbers to cause the fruit-grower any uneasiness. The beetle (Fig. 6) is

FIG. 6.

of rather an elegant form, with long, tapering antennæ of a gray color, prettily banded with black. It is a little more than a third of an inch long, of a brownish-gray color, with many small, thorn-like points upon its wing-covers. There is also a V-shaped band, margined with black, a little behind the middle of the wing-cases.

The perfect insect appears about the last of August, when it occasionally deposits its eggs upon the trunks of apple-trees, which shortly hatch into small grubs, and these eat their way through and burrow under the bark. They are very similar in appearance to the young larvæ of the two-striped borer, but differ in their habits; they form long, narrow, winding tracks under the bark, but upon the outer surface of the wood, which are made broader as the larva increases in size. This larva is also found under the bark of oak-trees.

Remedies.—Should the insect at any time prove destructive, its ravages may be prevented or controlled by the use of the alkaline wash applied to the bark, as recommended for the two-striped borer (No. 2), deferring its application until the early part of August.

No. 5.—The Stag Beetle.

Lucanus dama Thunb.

This large and powerful beetle is a very common insect, belonging to the family called *Lamellicornes*, or leaf-horned beetles, from the leaf-like joints of their antennæ. In the male (Fig. 7) the upper jaws or mandibles are largely de-

veloped, curved like a sickle, and furnished internally beyond the middle with a small tooth; those of the female are much shorter, and also toothed. The body measures from one to one and a quarter inches in length, exclusive of the jaws, and is of a deep mahogany-brown color. The head of the male is broad and smooth; that of the female narrowed and roughened with indentations. The beetle appears during the months of

Fig. 7.

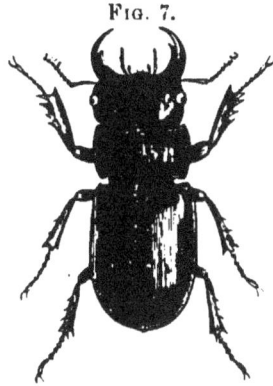

July and August, and is very vigorous on the wing, flying with a loud, buzzing sound during the evening and night, when it frequently enters houses, to the annoyance of the occupants. It is perhaps scarcely necessary to remark that this beetle is not venomous, and that it never attempts to bite without provocation.

The eggs are laid in the crevices of the bark of trees, especially near the roots. The larvæ live in decaying wood, and are found in the trunks and roots of various kinds of trees, particularly those of old apple-trees; they are also found in old cherry-trees, willows, and oaks. They are said to be six years in completing their growth, living all the time on the wood of the tree, reducing it to a coarse powder resembling sawdust. The mature larva is a large, thick, whitish grub, with a reddish-brown, horny-looking head, dark mandibles, and reddish legs. (See Fig. 8, *a.*) The body is curved when at rest, the hinder segments being brought towards the head.

FIG. 8.

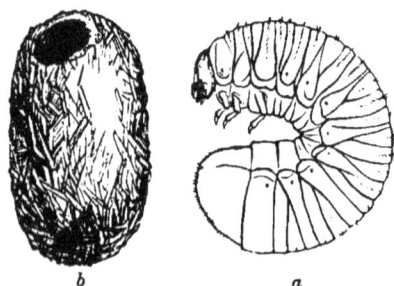

b   *a*

When the larva has attained full size it remains in its burrow, and encloses itself in an oval cocoon (Fig. 8, *b*) formed of fragments of wood and bark cemented together with a glue-like secretion, and within this enclosure it is transformed into a pupa of a yellowish-white color. Through the partially transparent membrane the limbs of the future beetle are dimly seen, and in due time the mature insect bursts its filmy covering, crawls through the passage previously gnawed by the larva, and emerges to the light of day.

As this beetle affects only old and decaying trees, it seldom does much harm. The use of the alkaline wash recommended for No. 2 would no doubt deter the beetles from depositing their eggs on trees so protected, and thus any mischief they might otherwise do could be prevented.

No. 6.—The Apple-bark Beetle.

Monarthrum mali (Fitch).

The apple-bark beetle is a small insect about one-tenth of an inch long (see Fig. 9, where it is shown much magnified);

it is cylindrical in form, smooth and slender, and varies in color from dark chestnut-brown to nearly black. Its legs and antennæ are pale-yellowish, and its thorax minutely punctated; the posterior end of the body is abruptly notched or excavated. The insect bores under the bark of apple-trees, sometimes attacking young, thrifty trees, which, when badly affected, are apt, soon after putting forth their leaves, to wither suddenly, as if scorched by fire; the bark becomes loosened from the wood, and soon after, these small beetles appear crawling through minute per-forations in the bark like large pin-holes. This insect usually appears in July; it is seldom very common, but has been reported as destructive in some parts of Massachusetts, where many young trees are said to have been ruined by it. So little is yet known of the history and habits of this pest that it is difficult to say what would be the best remedy for it.

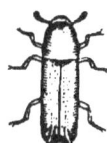

Fig. 9.

No. 7.—The Eyed Elater

Alaus oculatus (Linn.).

This is the largest of our Elaters, or "spring-beetles," and is found with its larva in the decaying wood of old apple-trees. The beetle (Fig. 10) is an inch and a half or more in length, of a black color, sprinkled with numerous whitish dots. On the thorax there are two large velvety black eye-like spots, which have given origin to the common name of the insect. The thorax is about one-third the length of the body, and is powdered with whitish atoms or scales; the wing-cases are ridged with longitudinal lines, and the under side of the body and legs thickly powdered with white. It is found in the perfect state in June and July.

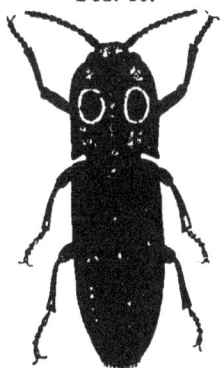

Fig. 10.

The mature larva (Fig. 11), which attains its full growth early in April, is about two and a half inches long, nearly four-tenths of an inch across about the middle, tapering

FIG. 11.

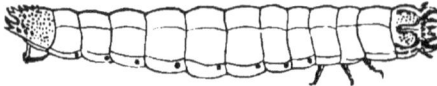

slightly towards each extremity. The head is broad, brownish, and rough above; the jaws very strong, curved, and pointed; the terminal segment of the body blackish, roughened with small pointed tubercles, with a deep semicircular notch at the end, and armed at the sides with small teeth, the two hindermost of which are long, forked, and curved upwards like hooks; under this hinder segment is a large fleshy foot, furnished behind with little claws, and around the sides with short spines; it has six true legs,—a pair under each of the first three segments. Early in spring the larva casts its skin and becomes a pupa, and in due time there emerges from it a perfect beetle.

This beetle, when placed upon its back on a flat surface, has the power of springing suddenly into the air, and, while moving, turning its body, thus recovering its natural position. This unusual movement combines with its curious prominent eye-like spots to make it a constant source of wonder and interest. Since it feeds mainly on decaying wood, it scarcely deserves to be classed with destructive insects; yet, being occasionally found in the trunk of the apple-tree, it is worthy of mention here.

No. 8.—The Rough Osmoderma.

Osmoderma scabra (Beauv.).

This insect, also, lives in the larval state in the decaying wood of the apple, as well as in that of the cherry, consuming the wood and inducing more rapid decay. It is a large, white, fleshy grub, with a reddish, hard-shelled head. In the autumn each larva makes for itself an oval cell of fragments of wood, cemented together with a glutinous material, in which it undergoes its transformations, appearing during the month of July as a large, purplish-black beetle (Fig. 12), about an inch long, with rough wing-cases. The head is hollowed out on the top, the under side of the body smooth, and the legs short and stout. It conceals itself during the day, but is active at night, feeding upon the sap which flows from the bark. Since the larva feeds chiefly on decaying wood, the injury inflicted, if any, can only be of a trifling character.

FIG. 12.

ATTACKING THE BRANCHES.

No. 9.—The Woolly-louse of the Apple.

Schizoneura lanigera (Hausm.).

This is the same species as the apple-root plant-louse (No. 1), but in this form the insects attack the trunk and limbs of the apple-tree, living in clusters, and secreting over themselves small patches of a cotton-like covering. (See Fig. 13, where the insects are represented magnified.) They are often found about the base of twigs or suckers springing from the trunk, and also about the base of the trunk itself, and around recent wounds in the bark. In autumn they commonly affect the axils of the leaf-stalks (Fig. 13), towards the ends of twigs, and sometimes multiply to such an extent as to cover the whole under surface of the limbs and also of the trunk, the tree looking as though whitewashed. They are said to affect most those trees which

FIG. 13.

yield sweet fruit. This woolly-louse is very common in Europe, especially in Germany, the north of France, and England, where it is more destructive than in this country, and, although generally known there under the name of the "American Blight," it is believed to be indigenous to Europe, and to have been originally brought from Europe to America. It appears to thrive only in comparatively cold climates, and in this country occurs in this form most abundantly in the New England States.

Under each of the little patches of down there is usually found one large female with her young. When fully grown the female is nearly one-tenth of an inch long, oval in form, with black head and feet, dusky legs and antennæ, and yellowish abdomen. She is covered with a white, mealy powder, and has a tuft of white down growing upon the hinder part of her back, which is easily detached. During the summer the insects are wingless, and the young are produced alive, but about the middle of October, among the wingless specimens, appear a considerable number with wings, and these have but little of the downy substance upon their bodies, which are nearly black and rather plump. The fore wings are

FIG. 14.

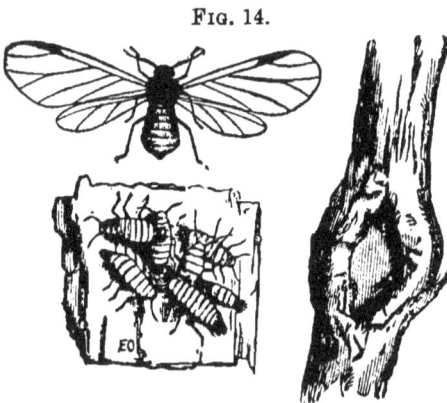

large, and about twice as long as the narrower hind wings. In Fig. 14 the winged insect is represented much magnified; also a group of the young lice magnified, and an apple-twig, natural size, showing one of the openings in the bark caused by this insect. Late in the autumn the females deposit eggs for another generation the following spring,—a fact which should induce fruit-growers to take particular pains to destroy these lice wherever found, for the colony that is permitted this year to establish itself upon some worthless tree or on the shoots or suckers at its base, will furnish the parents of countless hosts that may establish themselves next year on the choicest trees in the orchard. The insects are extremely hardy, and will endure a considerable amount of frost, and it is quite probable that some of them survive the winter in the perfect state in the cracks of the bark of the trees.

The eggs are so small that they require a magnifying-glass to enable one to see them, and are deposited in the crevices of the bark at or near the surface of the ground, especially about the base of suckers, where such are permitted to grow.

The young, when first hatched, are covered with very fine down, and appear in the spring of the year like little specks of mould on the trees. As the season advances, and the insect increases in size, its cottony coating becomes more distinct, the fibres increasing in length and apparently issuing from all the pores of the skin of the abdomen. This coating is very easily removed, adhering to the fingers when touched. Both young and old derive their nourishment from the sap of the tree, and the constant punctures they make give rise to warts and excrescences on the bark, and openings in it, and, where very numerous, the limbs attacked become sickly, the leaves turn yellow and drop off, and sometimes the tree dies.

Remedies.—The very small four-winged Chalcid fly, *Aphelinus mali* (Hald.), which is highly magnified in Fig. 15, and which has already been referred to under No. 1, preys also on this woolly aphis. The lady-birds and their larvæ, also the larvæ of the lace-wing flies and syrphus flies, feed on all species of plant-lice,

FIG. 15.

and are very useful in keeping them within bounds. These friendly insects will be fully treated of under the Apple-

tree Aphis, No. 57. The vigorous use of a stiff brush wet
with the alkaline solution of soap, recommended under
No. 2, will also be found very efficient, or a solution made
by mixing five pounds of fresh lime with one pound of
sulphur and two gallons of water, and heating until the
sulphur is dissolved. After destroying those on the trunk,
and cutting away all suckers, the earth should be removed
from about the base of the trunk, the parts below the surface
cleaned, and fresh earth placed about the roots. Spiders
devour large numbers of these lice, spinning their webs over
the colonies and feeding at their leisure.

No. 10.—The Apple Liopus.

Liopus facetus Say.

This is another of the long-horned borers which has been
found in the larval state boring into the decaying limbs of
apple-trees. The larva, when full grown, is a quarter of an
inch long or more, is slender, with the anterior segments enlarged and swollen, is covered with fine short hairs, and has
the end of the abdomen rather blunt. The beetle, which is
shown magnified in Fig. 16, is a handsome one, a slender

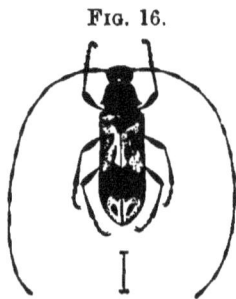

FIG. 16.

little creature, rather less than a quarter
of an inch in length, of a pale ash-gray
color with a purplish tinge. The long
antennæ are yellowish brown, except at
the base and between the joints, where
the color is darker. The wing-covers
are smooth, and on their anterior portion is an irregular rounded dark spot;
a broad black band crosses the hinder
portion, leaving the tip pale gray; there
are also several additional blackish dots and streaks distributed over the upper surface.

The beetles appear late in June and early in July, and lay
their eggs on the bark of the branches, from which the young
larvæ hatch and bore in under the bark, where they become

full grown and undergo their transformations before the following midsummer. This is a rare insect in most parts of America, and is not likely to prove a serious trouble anywhere.

No. 11.—The Apple-tree Pruner.

Elaphidion villosum (Fabr.).

This is also a long-horned beetle, of cylindrical form, of a dull-blackish color, with brownish wing-cases. The antennæ in the male are longer than the body, and in the female, which is represented in Fig. 17, are equal to it. The entire body is covered with short grayish hairs, which, from their denseness in some places on the thorax and wing-covers, form pale spots. The under side of the body is of a chestnut-brown color. The insect affects chiefly the oak-tree, but also attacks the apple, and, although not often found in great abundance, is very generally distributed over most of the Northern United States and Canada.

Fig. 17.

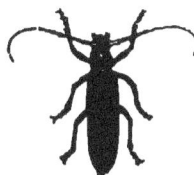

The peculiar habits and instincts of this insect are very interesting. The parent beetle places an egg in the axil of a leaf on a fresh green twig proceeding from a moderate-sized limb. When the young larva hatches, it burrows into the centre of the twig and down towards its base, consuming in its course the soft pulpy matter of which this part of the twig is composed. By the time it reaches the main limb it has become sufficiently matured to be able to feed upon the harder wood, and makes its way into the branch, when the hollow twig it has vacated gradually withers and drops off. The larva, being now about half grown, eats its way a short distance through the middle of the branch, and then proceeds deliberately to sever its connection with the tree by gnawing away the woody fibre to such an extent that the first storm of wind snaps the branch off. This is rather a delicate operation for the insect to perform, and requires wonderful instinctive skill, for should it gnaw away too much of the woody interior the branch might break during the process, —an accident which would probably crush the workman to death; but the insect rarely miscalculates: it leaves the bark and just enough of the woody fibre untouched to sustain the branch until it has time to make good its retreat into the burrow, the opening of which it carefully stops up with gnawed fragments of wood. If the limb be short, it severs all the woody fibres, leaving it fastened only by the bark; if longer, a few of the woody fibres on the upper side are left; and if very long and heavy, not more than three-fourths of the wood will be cut through. Having performed the operation and closed its hole so that the jarring of the branch when it falls may not shake out the occupant, the larva retreats to the spot at which it first entered the limb. After the branch has fallen it eats its way gradually through the centre of the limb for a distance of from six to twelve inches,

Fig. 18.   Fig. 19.

when, having completed its growth, it is transformed to a pupa within the enclosure. Sometimes this change takes place in the autumn, but more frequently it is deferred until the spring, and from the pupa the beetle escapes during the month of June.

The larva (Fig. 18) when full grown is a little more than half an inch long, thickest towards the head, tapering gradually backwards. The head is small and black, the body yellowish white, with a few indistinct darker markings. It has six very minute legs attached to the anterior segments. In the figure the larva is shown magnified. The pupa is about the same size as the larva, of a whitish color, and is shown in Fig. 19, also magnified, in its burrow.

Remedies.—Birds are active agents in the destruction of these larvæ; they seek them out in their places of retreat and devour them. Should they at any time become very numerous, they may easily be disposed of by gathering the fallen branches and burning them before the insect has time to mature.

No. 12.—The Parallel Elaphidion.

Elaphidion parallelum Newm.

This insect in the larval state occasionally bores into the twigs of apple and plum trees. The beetle (Fig. 20, *c*) is a little more than half an inch long, of a dull-brownish color, closely resembling No. 11 in appearance and habits, but smaller in size.

The egg is laid by the parent insect near the axil of one of the leaf-buds, where the young larva, when hatched, bores into the twig, enlarging the channel as it increases in size, finally transforming to a pupa within its burrow, and escaping at maturity in the perfect state. In the figure, *a* shows the larva, *b* the twig split open, showing the enclosed pupa, *k* the end of the twig cut off, *c* the beetle; *i* the basal joints of the antenna, *j* the tip of the wing-case, *d* the head, *e* maxilla, *f* labium, *g* mandible, and *h* the antenna of the larva. This Elaphidion is rather a rare insect, and, although it may

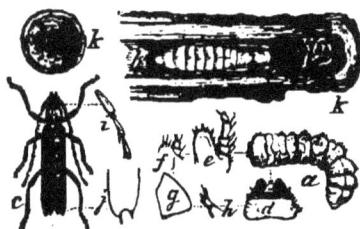

Fig. 20.

occasionally be found injurious, it is not likely to become so to any considerable extent.

No. 13.—The Apple-twig Borer.

Amphicerus bicaudatus (Say).

The apple-twig borer is a small cylindrical beetle (Fig. 21), from one-fourth to one-third of an inch in length, of a dark chestnut-brown color above, black beneath. The fore part of its thorax is roughened with minute elevated points, and, in the males, furnished with two little horns ; the male may also be further distinguished from the female by its having two small thorn-like projections from the extremities of the wing-covers.

Unlike most other borers, which do their mischief in the larval state, this insect works in the beetle state, boring into the branches of apple, pear, and cherry trees, just above a

FIG. 21.   FIG. 22.

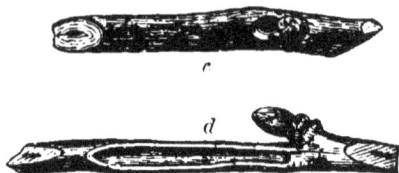

bud, and working downwards through the pith in a cylindrical burrow one or two inches long. (See Fig. 22, *c* and *d*.) The holes appear to be made partly for the purpose of obtaining food, and partly to serve as places of concealment for the beetles ; they are made by both sexes alike, and the beetles are found in them occasionally in the middle of winter, as well as in the summer, usually with the head downwards. They work throughout the summer months, causing the twigs operated on to wither and their leaves to turn brown. Upon examination, a perforation about the size of a knitting-needle is found near one of the buds from six inches to a foot from the end of the twig. This insect does not often occur in such numbers as to inflict any material damage, but occasionally as many as ten have been found working at once on a two- or three-year-old tree ; they also affect the twigs of larger trees. The twigs so injured are very liable to break off with high winds.

There is not much known as yet about the earlier stages of this insect ; the larva is said to have been found feeding upon grape-canes, into which also the beetle occasionally bores. The beetle is found from Pennsylvania to Mississippi, also in the orchards of New Jersey, Michigan, Illinois, Iowa, and Kansas. Should it at any time inflict serious injury, the only remedy as yet suggested is to search for the bored twigs in June and July, and cut them off and burn them.

No. 14.—The Imbricated Snout-beetle.

Epicœrus imbricatus (Say).

This is a small snout-beetle or weevil, which is common in some localities on apple and cherry trees and injures them by gnawing the twigs and fruit. It is most frequently found in the Western States, especially in parts of Iowa and Kansas.

Fig. 23.

It is a very variable beetle; usually it is of a silvery-white color, with dark markings, as shown in Fig. 23, but sometimes these latter are wholly or partly wanting. Nothing is as yet known of its history in the earlier stages of its existence.

Should this weevil ever occur in sufficient numbers to excite alarm, they could probably be collected by jarring the trees, as in the case of the plum-weevil, and then destroyed.

No. 15.—The Seventeen-year Locust.

Cicada septendecim Linn.

The seventeen-year locust is an insect very well known throughout the United States, and is sometimes met with in Canada. It is generally believed to require seventeen years in which to complete its transformations, nearly the whole of this period being spent under ground.

The perfect insect measures, when its wings are expanded, from two and a half to three inches across. It is represented at *c* in Fig. 24. The body is stout and blackish, the wings transparent, the thick anterior edge and large veins are orange-red, and near the front margin, towards the tip, there is a dusky, zigzag line resembling a W. The rings of the abdomen are edged with dull orange, and the legs are of the same hue. The locusts appear in the South earlier than in

FIG. 24.

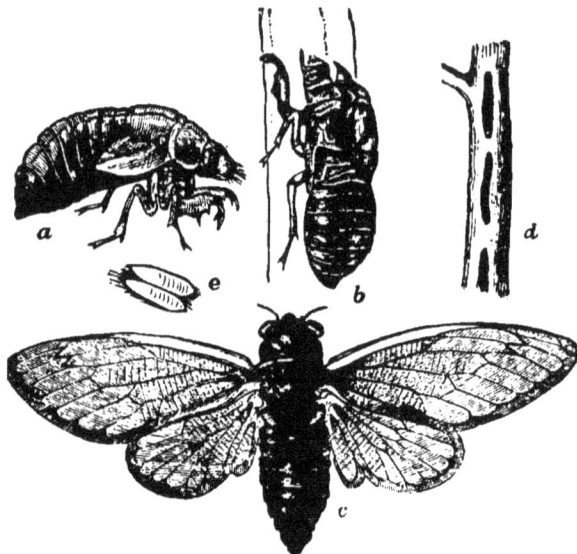

the North; their usual time is during the latter part of May, and they disappear early in July.

After pairing, the female deposits her eggs in the twigs of different trees, puncturing and sawing small slits in them, as shown in Fig. 24, *d*, which she does by means of her sharp beak, which is composed of three portions; the two outer are beset with small teeth like a saw, while the centre one is a spear-pointed piercer. In these slits she places her eggs. These (*e*, Fig. 24) are of a pearly-white color, one-twelfth of an inch long, and taper to an obtuse point at each end. They are deposited in pairs, side by side, with a portion of woody fibre between them, and placed in the cavity some-what obliquely, so that one end points upwards. When two eggs have thus been deposited, the insect withdraws her piercer for a moment, and then inserts it again and drops two more eggs in a line with the first, and so on until she has filled the slit from one end to the other. She then removes to a little distance and makes another similar nest: it is not uncommon to find from fifteen to twenty of such fissures in the same limb. The cicada thus passes from limb to limb and from tree to tree until her store of four or five hundred eggs is exhausted, when, worn out by her excessive labors, she dies. The punctured twigs are so weakened by the operations of the insect that they frequently break off when swayed by rough winds, and the injury thus caused to young fruit-trees in orchards or nurseries is sometimes very serious ; in most instances, however, if the trees are vigorous, they eventually recover from their wounds.

The eggs hatch in about six weeks or less, the young larva being of a yellowish-white color, and appearing as shown in Fig. 25. It is active and rapid in its movements, and shortly after its escape from the egg drops to the ground, and immediately proceeds to bury itself in the soil by means of its broad and strong fore feet, which are admirably adapted for digging. Once under the surface, these larvæ attach themselves to the succulent roots of plants and trees, and, puncturing them with their beaks, imbibe the vegetable juices, which form their sole nourishment. They do not usually descend very deeply

Fig. 25.

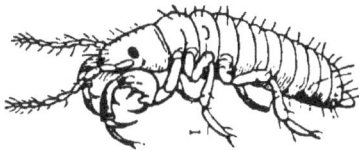

into the ground, but remain where juicy roots are most abundant, and the only marked alteration to which they are subject during the long period of their existence under ground is a gradual increase in size.

As the time for their transformation approaches, they ascend towards the surface, making cylindrical burrows about five-eighths of an inch in diameter, often circuitous, seldom exactly perpendicular, and these are firmly cemented and varnished so as to be water-tight. As the insect progresses, the chamber is filled below by the earthy matter removed in its progress, but the upper portion, to the extent of six or eight inches, is empty, and serves as a dwelling-place for the insect until the period for its exit arrives. Here it remains for some days, ascending to the top of the hole in fine weather for warmth and air, and occasionally looking out as if to reconnoitre, but descending again on the occurrence of cold or wet weather. In localities that are low or imperfectly drained, the insects sometimes continue their galleries from four to six inches above ground, as shown in Fig. 26, leaving a place of egress at the surface, e, and in the upper end of these dry chambers the pupæ patiently await the time for their next change.

Fig. 26.

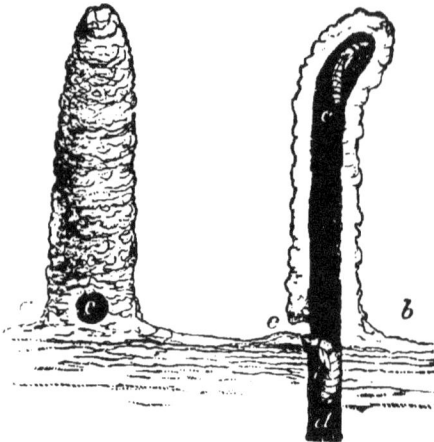

This period, although an active one, is the pupal stage of the insects' existence, and finally, when fully matured, they issue from the ground (see a, Fig. 24), crawl up the trunk of a tree or any other object to which they can attach themselves securely by their claws, and, having rested awhile, prepare to cast their skins. After some struggling, a longitudinal rent is made on the back, and through this the enclosed cicada pushes its head, and then gradually withdraws itself, leaving the empty pupa skin adhering, as shown at b in Fig. 24. The escape from the pupa usually occurs between six and nine in the evening, and about ten minutes are occupied by the insect in entirely freeing itself from the enclosure. At first the body is soft and white, excepting a black patch on the back, and the wings are small and soft, but within an hour are fully developed, and before morning the mature insects are ready for flight. They sometimes issue from the ground in immense numbers; above fifteen hundred have been known to arise beneath a single apple-tree, and in some places the whole surface of the soil has, by their operations, appeared almost as full of holes as a honey-comb.

Remedies.—On escaping from the ground, they are attacked by various enemies. Birds and predaceous insects devour them; hogs and poultry feed on them greedily; and in the winged state they are also subject to the attacks of parasites. It seems that human agency can effect but little in the way of staying the progress of these invaders, and the only time when anything can be done is early in the morning, when the winged insects newly escaped and in a comparatively feeble and helpless condition may be crushed and destroyed; but when once they have acquired their full power of wing, it is a hopeless task to attempt to arrest their course. The males have a musical apparatus on each side of the body just behind the wings, which acts like a pair of kettle-drums, producing a very loud, shrill sound. Although partial to oak-trees, on which they most abound, they are very destructive to other trees and shrubs, and frequently injure apple-trees.

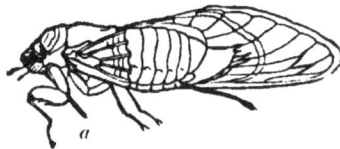

FIG. 27.

A popular idea prevails that these insects are dangerous to handle, that they sting, and that their sting is venomous. As their beaks (*a*, Fig. 27) are sharp and strong, it is possible that under provocation they may insert these, but, since there is no poison-gland attached, there is little more to fear from their puncture than from the piercing of a needle.

No. 16.—The Oyster-shell Bark-louse.

Mytilaspis pomorum Bouché.

This is a very destructive and pernicious insect, which prevails throughout the Northern United States and Canada, and in some of the Southern States also. It was introduced from Europe more than eighty years ago. It appears in the form of minute scales, about one-sixth of an inch long, of a brownish or grayish color, closely resembling that of the bark of the tree, and somewhat like the shell of an oyster in shape, adhering to the surface of the bark, as shown in Fig. 28, and placed irregularly, most of them lengthwise of the limb or twig, with the smaller end upwards. In some instances the branches of apple-trees may be found literally covered and crowded with these scales; and where thus so prevalent they seriously impair the health and vigor of the tree, and sometimes cause its death.

Fig. 28.

Under each of these scales will be found a mass of eggs varying in number from fifteen or twenty to one hundred or more; these during the winter or early spring will be found to be white in color, but before hatching they change to a yellowish hue, soon after which the young insects appear. This usually occurs late in May or early in June, and, if the weather is cool, the young lice will remain several days under the scales before dispersing over the tree. As it becomes warmer, they leave their shelter, and may be seen running all over the twigs looking for suitable locations to which to attach themselves. They then, under a magnifying-glass, present the appearance shown at 2, Fig. 29, their actual length being only about one-hundredth of an inch; to the unaided eye they appear as mere specks. A large proportion of them soon become fixed around the base of the side-shoots of the terminal twigs, where, inserting their tiny sharp beaks, they subsist upon the sap of the tree. In a few days a fringe of delicate waxy threads issues from their bodies, when they have the appearance shown at 3. Gradually the insect assumes the form shown at 4; 5 and 6 represent the louse as it approaches maturity, and when detached from the scale; 1 shows the egg highly magnified; and 8 one of the antennæ of the young lice, also much enlarged. Before the end of the season the louse has secreted for itself

FIG. 29.

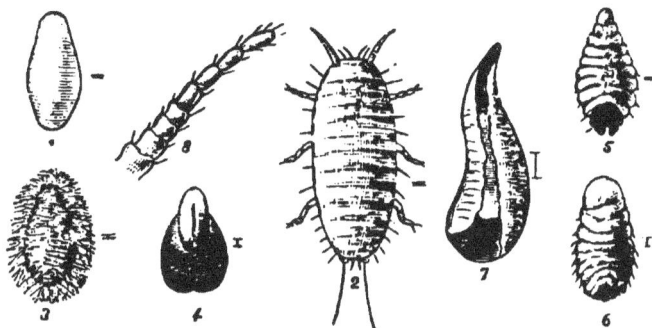

the scaly covering shown at 7, in which it lives and matures. The scale is figured as it appears from the under side when raised and with the louse in it. By the middle of August this female louse has become little else than a bag of eggs, and the process of depositing these now begins, the body of the parent shrinking day by day, until finally, when this work is completed, it becomes a mere atom at the narrow end of the scale, and is scarcely noticeable.

The scales of the male louse are seldom seen; they are most frequently found upon the leaves, both on the upper and under sides; they are smaller in size than those of the female, and different also in shape. The male scale is shown at *c*, Fig. 30, in which cut is also represented the male insect, much magnified, with wings closed and expanded.

Only one brood is produced annually in the North, the eggs remaining unchanged under the scale for about nine months; but in some parts of the South the insect is double-brooded, the first brood hatching in May, the second in September.

As the oyster-shell bark-louse retains power of motion only for a few days at most after hatching, it is mainly disseminated to distant places by the distribution of young trees from infested nurseries. In the orchard and its immediate neighborhood it may be spread by being carried on the feet of birds, or attached

Fig. 30.

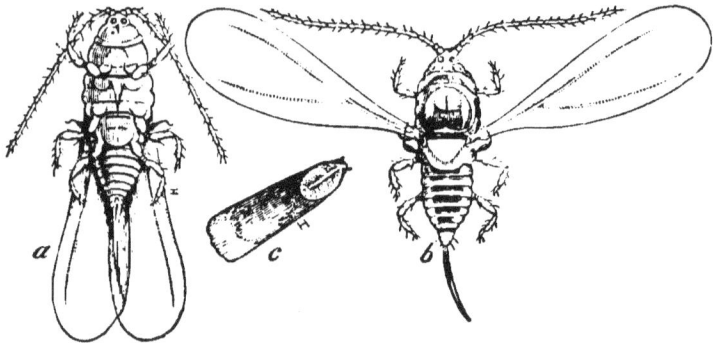

to the larger insects, or may be aided by the wind in passing from tree to tree, while it is itself so brisk in its active state that it can travel two or three inches in a minute, and hence might in this way reach a point two or three rods distant before it would perish. Although this insect essentially belongs to the apple-tree, it is frequently found on the pear, and sometimes on the plum.

Remedies.—A species of mite (Fig. 31), *Tyroglyphus malus* (Shimer), preys on the louse as well as on its eggs; and this mite, so insignificant that it can scarcely be seen without a magnifying-glass, has probably done more to keep this orchard-pest within bounds than any other thing.

Under the scales may sometimes be found a small active larva devouring the eggs. This is the progeny of a small four-winged parasite, belonging to the family Chalcididæ, named *Aphelinus mytilaspidis* Le Baron. In Fig. 32 we have a representation of this insect highly magnified.

Another friend is the twice-stabbed lady-bird, *Chilochorus bivulnerus* Muls. (Fig. 33), an insect easily recognized by its

FIG. 31.

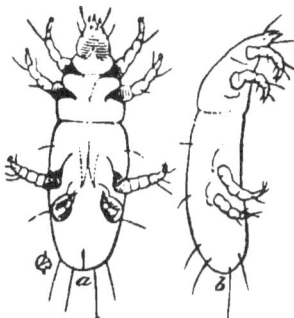

FIG. 32.

polished black wing-cases with a blood-red spot on each. Its larva, a bristly-looking little creature (Fig. 34), of a grayish color, is very active, and devours large numbers of the lice; the perfect beetle also eats them. The bark-lice and their eggs are devoured also by some of our insect-eating birds.

FIG. 33.

During the winter the trees should be examined and the scales scraped off, and thus a large proportion of the insects may be destroyed. Still, it is almost impossible to cleanse the trees entirely in this way, especially the smaller branches; and hence the insect should be fought also at the time when the eggs are hatching and the young lice crawling over the limbs, as then they are tender and easily killed. With this object in view, the time of hatching of the remnants left after the winter or spring scraping should be watched, and, while the young larvæ are active, the twigs should be brushed with a strong solution of soft-soap and washing-soda, as recommended under No. 2, or syringed with a solution of washing-soda in water, made by dissolving half a pound or more

FIG. 34.

in a pailful. Painting the twigs and branches with linseed oil has also been tried with success.

As a precautionary measure, every young tree should be carefully examined before being planted, and if found infested should be thoroughly cleansed.

No. 17.—The Scurfy Bark-louse.
Chionaspis furfurus (Fitch).

This insect, which has long been known under the name of Harris's Bark-louse, *Aspidiotus Harrisii* Walsh, is now found to have been first described by Dr. Fitch, and hence must in future bear the name given to it by him. It resembles in some respects the oyster-shell bark-louse, yet is sufficiently dissimilar to be readily distinguished from it. In this species the scale of the female, which is by far the most abundant, is oblong in form, pointed below, very flat, of a grayish-white color, and about one-tenth of an inch long. (See Fig. 35, 1 and 1 *c;* the latter represents a scale highly magnified.) The eggs under the scale of the oyster-shell bark-louse during the winter are white, while these are purplish red. The eggs of this species hatch about the same date as the other, but the larvæ are red or reddish brown in color. This insect does not mature so rapidly as the oyster-shell species; the eggs are said not to be fully developed under the scale until the middle of September. The scale of the male, which is very much smaller and narrower, and not more than one-thirtieth of an inch long, is shown in the figure, magnified, at 1 *a;* the male insect in the winged state, highly magnified, at 1 *b.*

This is a native insect, which has existed from time immemorial in the East, West, and South, its original home being on the bark of our native crab-trees. In the warmer parts of the South it is more common than the oyster-shell bark-louse. It is found chiefly on the apple, but sometimes affects the pear and also the mountain-ash. It is far less common than the imported oyster-shell bark-louse, and is nowhere anything like so injurious as that insect.

Remedies.—The scurfy bark-louse is said to be preyed upon by the same mites which attack the oyster-shell species; it is

FIG. 35.

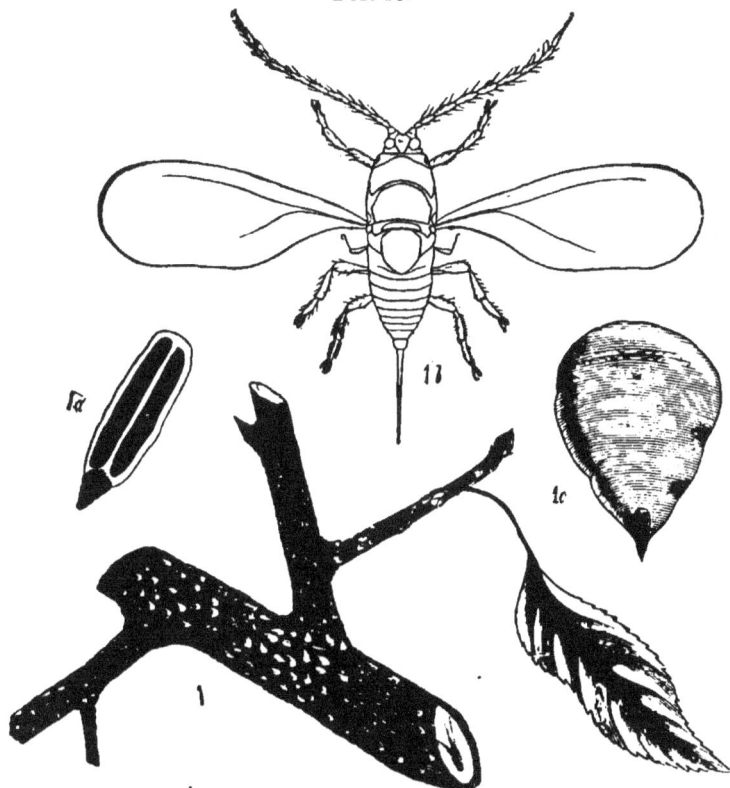

also devoured by the larva of the twice-stabbed lady-bird. The same artificial remedies should be used in this instance as are recommended in the other.

No. 18 —The Buffalo Tree-hopper.

Ceresa bubalus (Fabr.).

This insect belongs to the order Hemiptera. It is an active jumping creature, about one-third of an inch long (Fig. 36), of a light grass-green color, with whitish dots and a pale-yellowish streak along each side. On the front there is a sharp process or point jutting out horizontally on each side, reminding one of the horns of a bull or buffalo, which has given to the insect its common name of buffalo tree-hopper. Its body is three-sided, not unlike a beech-nut in form, and it is furnished with a sharp-pointed beak, with which it punctures the bark and sucks the sap from the trees.

Fig 36.

It is common on apple and many other trees from July until the end of the season.

The eggs are said to be laid in a single row of slits in the bark, and when hatched the young larvæ, which are grass-green like their parents, feed also on the sap of the leaves and twigs.

In the larval state, before the power of flight is acquired, the insect is easily caught and destroyed; but it is not easy to suggest a remedy for so active a creature as the perfect insect is. It cannot be killed by any poisonous application, as it feeds only on sap. It has been suggested that where they are so numerous as to injure fruit-trees they may be frightened away by frequently shaking the trees, as they are very shy and timorous. It is, however, scarcely probable that this insect will ever become a source of much annoyance to the fruit-grower.

No. 19.—The Thorn-bush Tree-hopper.

Thelia cratœgi Fitch.

Fig. 37.

This is an insect similar in structure and habits to the buffalo tree-hopper. It is common on apple-trees, but more common on thorn-bushes, in July and August, when it may be seen resting upon the small limbs and sucking the sap. When approached, it leaps away with a sudden spring, and is lost to view.

It is a little more than one-third of an inch long (see Fig. 37), with a three-sided body, black, varied with chestnut-brown, with a large white spot on each side, which extended forward becomes a band across the front. There is also a white band across the hind part of its back, and a protuberance extending upwards on the front part of its body.

ATTACKING THE LEAVES.

No. 20.—The Apple-tree Tent-caterpillar.
Clisiocampa Americana Harris.

This insect is a native of the more northern Atlantic States, and has probably been carried westward in the egg-state attached to the twigs of young trees. It inhabits now almost

Fig. 38.   Fig. 39.

all parts of the United States and Canada. The moth is of a pale dull-reddish or reddish-brown color, crossed by two oblique parallel whitish lines, the space between these lines being usually paler than the general color, although sometimes quite as dark, or darker. In the male (Fig. 38) the antennæ are pectinate, or feather-like, and slightly so in the female (Fig. 39). When fully expanded, the wings of the female will measure an inch and a half or more across; the male is smaller. The hollow tongue or tube by which moths and butterflies imbibe their food is entirely wanting in this species; hence it has no power of taking food, and lives but a very few days in the winged state, merely long enough to provide for a future generation by the deposition of eggs. The moth remains at rest and concealed during the day, but becomes very active at night, when it enters lighted rooms, attracted by the glare, and becomes so dazzled and bewildered that it darts crazily about, here and there, thumping itself against the walls, furniture, and floor of the room in the most erratic manner, then circles around the lamp or gas-light with great velocity, finally dashing into the flame, when, with wings and antennæ severely singed, it retreats into some obscure corner. The moths are most abundant during the first two weeks in July.

The eggs are deposited during that month upon

Fig. 40.

the smaller twigs of our fruit-trees in ring-like clusters, each composed of from fifteen to twenty rows, containing in all from two to three hundred. The eggs are conical and about one-twentieth of an inch long, firmly cemented together, and coated with a tough varnish, impervious to rain, the clusters presenting the appearance shown in Fig. 40. In Fig. 41, at *c*, a similar cluster is shown with the gummy covering removed, showing the manner in which the eggs are arranged.

The young caterpillars are fully matured in the egg before winter comes, and they remain in this enclosure in a torpid state throughout the cold weather, hatching during the first warm days of spring. They usually appear during the last week in April or early in May, depending much on the prevailing temperature. Their first meal is made of portions of the gummy material with which the egg-masses are covered, and with the strength thus gained they proceed at once to work. At this time the buds are bursting, thus providing these young larvæ with an abundance of suitable tender food. It sometimes happens, however, that after they are hatched cold weather returns and vegetable growth is temporarily arrested. To meet this emergency they have the power of sustaining hunger for a considerable time, and will usually

live from ten to twelve days when wholly deprived of food ;
but severe frost is fatal to them in this tender condition, and
multitudes of them sometimes perish from this cause. These
larvæ are tent-makers, and soon after birth they begin to construct for themselves a shelter by extending sheets of web
across the nearest fork of the twig upon which they were

Fig. 41.

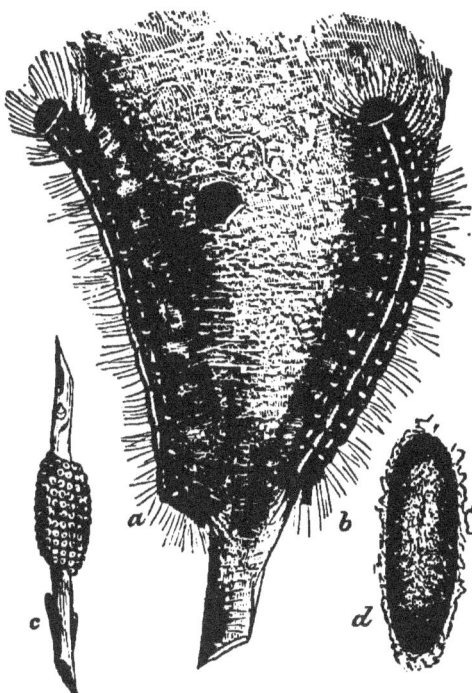

hatched. As they increase in size, they construct additional
layers of silk over those previously made, attaching them to
the neighboring twigs, and leaving between the layers space
enough for the caterpillars to pass. The tent or nest when
completed is irregular in form, about eight or ten inches in
diameter, and the holes through which the caterpillars enter
are situated near the extremities or angles of the nest, and into
this they retreat at night or in stormy weather, also at other

4

times when not feeding. In five or six weeks they become full grown, and then measure about an inch and three-quarters in length, and present the appearance shown in Fig. 41. The body is hairy and black, with a white stripe down the back, and on each side of this central stripe there are a number of short, irregular, longitudinal yellow lines. On the sides are paler lines, with spots and streaks of pale blue. The under side of the body is nearly black.

These caterpillars have regular times for feeding, issuing from the openings in their tent in processional order, usually once in the forenoon and once in the afternoon. In very warm weather they sometimes repose upon the outside of the nest, literally covering it and making it appear quite black with their bodies. They are very voracious, and devour the leaves of the trees they are on with great rapidity; it is estimated that each larva when approaching maturity will consume two leaves in a day, so that every day that a nest of such marauders is permitted to remain on a tree there is a sacrifice of about five hundred leaves. Where there happen to be several nests on one tree, or if the tree itself is small, they often strip every vestige of foliage from it, and in neglected orchards the trees are sometimes seen as bare of leaves in June as they are in midwinter. As the caterpillars arrive at maturity they leave the trees and wander about in all directions in search of suitable places in which to hide during their chrysalis stage. A favorite place is the angle formed by the projection of the cap-boards of fences or fence-posts.

Here they construct oblong oval cocoons (Fig. 41, *d*) of a yellow color, formed of a double web, the outer one loosely woven and slight in texture, the inner one tough and thick. In its construction the silk is mixed with a pasty substance, which, when dry, becomes powdery and resembles sulphur in appearance. Within these cocoons the larvæ change to brown chrysalids, from which, in about two or three weeks, the moths escape. This insect feeds on many different trees, but is particularly fond of the apple and wild cherry.

Remedies.—Since the tent-caterpillar is so easily detected
by its conspicuous nest, it need never become very troublesome, as the larvæ may be easily destroyed while sheltering
within it. They seldom leave the nest to feed until after 9
A.M., and usually return before sundown ; hence the early and
late hours of the day are the best times for destroying them.
With a suitable ladder and a gloved hand the living mass
may be seized and crushed in a moment, or the nest may be
torn from the tree and trampled under foot. Where a ladder
is not at hand, the nests may be removed by a pole with a
bunch of rags tied around the end of it. This work is most
easily done while the larvæ are young, and should be attended to as soon as the cobweb-like nests can be seen. Sometimes when the nest is destroyed a portion of the caterpillars
will be absent feeding, and within a few days it may be found
partly repaired, with the remnants of the host within it : so
that to subdue them entirely repeated visits to the orchard
should be made, and not a fragment of a nest permitted to
remain. Governments might well enforce under penalties the
destruction of these caterpillars, as their nests are so conspicuous that there can be no excuse for neglecting to destroy
them, and it is unfair that a careful and vigilant fruit-grower
should be compelled to suffer from year to year from the
neglect of a careless or indolent neighbor. Neglected trees
are soon stripped of their leaves, and become prematurely
exhausted by having to reproduce at an unseasonable time
their lost foliage ; with fruit-trees this is so great a tax on
their vital powers that they usually bear little or no fruit the
following season. The egg-clusters may be sought for and
destroyed during the winter months, when, the trees being
leafless, a practised eye will readily detect them. A cloudy
day should be selected for this purpose, to avoid the inconvenience of too much glare from the sky.

Several parasites attack this insect. A minute Ichneumon
fly, about one-twenty-fifth of an inch in length, is parasitic on
the eggs. By means of a long ovipositor it bores through

the outer gummy covering and egg-shell, and deposits its eggs within the egg of the tent-caterpillar, where the young larvæ of the parasite hatch and feed upon the contents of the egg-shell of our enemy. A small mite, very similar to that shown in Fig. 31, is also very destructive to these eggs, eating into them and feeding on their occupants. Two larger Ichneumon flies prey upon the caterpillar, *Pimpla conquisitor* (Say) (Fig. 42) and *Ichneumon lætus* Brullé, as well as one or more species of Tachina flies, two-winged insects a little larger than the common house-fly, similar to Fig. 46. All these latter parasites watch their opportunity when the growing caterpillar is feeding, and deposit their eggs on or under the skin of their victim, which shortly hatch, when the larvæ burrow into the bodies of the tent-caterpillars and feed on them, carefully avoiding the destruction of the vital organs. The infested caterpillars usually reach maturity and construct their cocoons, but after a time, instead of the moth, one or more of these friendly insects make their appearance. Several predaceous insects also devour the caterpillars; these are referred to in detail under No. 21.

FIG. 42.

No. 21.—The Forest Tent-caterpillar.

Clisiocampa sylvatica Harris.

This insect closely resembles the common tent-caterpillar, No. 20. The moth (*b*, Fig. 43) is of a similar color, but paler, or more yellowish. The space between the two oblique lines is usually darker than the rest of the wing, and the lines themselves are dark brown instead of whitish. In the figure, *a* represents the egg-cluster, *c* one of the eggs, much enlarged, as seen from the top, *d* a side-view of the same.

The eggs of this species may be distinguished by their almost uniform diameter and by their being cut off squarely at each end. The number of eggs in each cluster is usually from three to four hundred ; they are white, about one-twenty-fifth of an inch long, and one-fortieth wide, rounded at the base, gradually enlarging towards the apex, where they are margined by a prominent rim, and have a sunken spot in the centre. The eggs are deposited in circles, and with each one is secreted a small quantity of gummy matter, which firmly

Fig. 43.

fastens it to the twig and also to the adjoining egg, and upon becoming dry forms a coating of brown varnish over the pale egg. Like the tent-caterpillar, the young become fully formed in the eggs before winter, and remain within them in a torpid condition until spring.

The larvæ in this instance also hatch about the time of the bursting of the buds, and in the absence of food are endowed with similar powers of endurance. It is said they have been known to survive a fast of three weeks' duration. While young, they spin a slight web or tent against the side of the trunk or branches of the tree on which they are situated, but, from its peculiar color or slight texture, it is seldom noticed. In this early stage they often manifest strange processionary habits, marching about in single or double column, one larva so immediately following another that when thus crossing a sidewalk or other smooth surface they appear at a little distance like black streaks or pieces of black cord stretched across it. From the time they are half grown, until they approach maturity, they seem to have a great fondness for exercise, and delight to travel in rows along fence-boards, which they do at a very brisk pace when in search of food.

In about six weeks this larva becomes full grown (Fig. 44), and is then an inch and a half or more in length, of a pale-bluish color, sprinkled all over with black points and dots. On the back is a row of ten or eleven oval or diamond-

Fig. 44.

shaped white spots, by which it may be at once distinguished from the common tent-caterpillar, while on the sides there are pale-yellowish stripes, somewhat broken, and mixed with gray. The hairs on the body are fox-colored, mixed with coarser whitish hairs. The caterpillars attain full growth about the middle of June.

Occasionally, during the latter part of May, when about half grown and extremely voracious, these caterpillars will appear in perfect swarms and attract general attention. During the latter part of the day, and frequently also in the morning, they collect on the trunks and larger branches of the trees in large black masses, which are so easily reached that they seem to invite destruction. While particularly injurious to the apple, they also attack various species of forest-trees, such as oak, thorn, ash, basswood, beech, plum, cherry, walnut, hickory, etc., and sometimes large clumps of wood may be seen in June quite bare of foliage from the devastation caused by this insect, while underneath the ground is covered with small black grains of exuvia. It is often very abundant in the West, and occasionally equally destructive in the South, especially in Georgia and Tennessee.

When full grown, this larva spins a cocoon (see Fig. 45) closely resembling that of the tent-caterpillar, usually within the shelter of a leaf, the edges of which are partly drawn together. Within such an enclosure there is generally one cocoon, but in times of great abundance, and where the enclosure is large enough, there are often two or three cocoons together. At such periods almost every leaf or fragment of a leaf is so occupied, and, the whitish-yellow cocoons being only partly hidden, and the leaves hanging with their weight, one is impressed with the idea that the tree is laden with some strange sort of fruit. If leaves cannot be had for shelter, the cocoons will be found under the bark of trees, in every suit-able crevice or hiding-place in fences, or under logs. In two or three days the enclosed larva changes to a chrysalis of a red-dish-brown color, densely clothed with short pale-yellowish hair, and in the course of two or three weeks the moth appears, which, like the insect last described, No. 20, is nocturnal in its habits, and lives but a few days, when, having provided for the contin-uance of its species, it perishes.

Remedies.—The egg-clusters should be sought for and de-stroyed during the winter months. When the caterpillars are young, they will drop, sus-pended by a silken thread, in mid-air, if the branch on which they are feeding be suddenly struck ; advantage may be taken of this habit, and by swinging a stick around, the threads may be gathered in with the larvæ attached to them. When the caterpillars have become half grown, the trees should be frequently inspected, early in the morning, and the congregated masses crushed and destroyed with a stiff broom or some other equally suitable implement. During the day they are so constantly on the move, that a young tree thoroughly cleansed from them in the morning may be crowded again before evening. To avoid the necessity of constant watch-ing, strips of cotton batting, three or four inches wide, should be tied around the tree about half-way up the trunk ; these

Fig. 45.

bands should be tied tightly in the middle.   Each caterpillar
is furnished with four pairs of fleshy prolegs, which are
fringed with small horny hooks, and on its trying to pass
over the cotton these hooks get so entangled in the fibres
that further progress becomes very difficult, and is seldom
persisted in.   A shower of rain will pack the fibres of the
cotton somewhat, but where the string fastening it is tied
around the middle, the upper half washes down and makes
a sort of roof overhanging the lower portion, which in great
measure protects it from the weather.

Fig. 46.

These larvæ are seldom abundant for many years in succession,
for in times of great plenty their
natural enemies multiply with
amazing rapidity.   Several parasites destroy them.   Two species
of Ichneumon flies prey on them,
also a two-winged Tachina fly,
closely resembling the Red-tailed
Tachina fly, *Nemoræa leucaniæ* (Kirkp.) (Fig. 46), which
attacks the army-worm, but this fly is without the red tail.

Fig. 47.

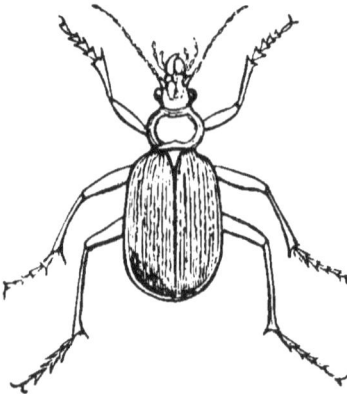

Fig. 48.

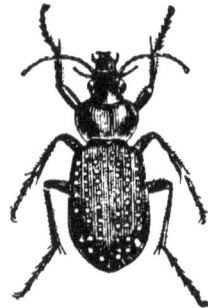

A species of bug (Hemiptera) attacks the caterpillars just when
they are constructing their cocoons, and sucks them empty,

while some of the insect-feeding birds devour them greedily,
especially the black-billed cuckoo. There are several species
of predaceous insects belonging to the *Carabidæ*, or ground-
beetles, which are very active in their habits, and diligently
hunt for them and eat them, notably the Green Caterpillar-
hunter, *Calosoma scrutator* (Fabr.) (Fig. 47), and the Copper-
spotted Calosoma, *Calosoma calidum* (Fabr.) (Fig. 48). They
are sometimes destroyed in great numbers by a fungoid disease,
which arrests their progress when
about full grown, and the affected
specimens may be found attached to
fences and trees, retaining an appearance almost natural, but when
handled they will often be found so
much decayed as to burst with a
gentle touch. An Ichneumon fly,
Pimpla pedalis Cresson (Fig. 49), is a parasite on this larva,
while mites prey upon the eggs, identical with those which
feed on the eggs of the common tent-caterpillar.

Fig. 49.

No. 22.—The White-marked Tussock-moth.

Orgyia leucostigma (Sm. & Abb.).

The orchardist, walking among his fruit-trees after the
leaves have fallen, or during the winter months, will frequently find a dead leaf or leaves fastened here and there to
the branches of his trees ; on examination, these will usually
be found to contain a gray cocoon, with in most instances a
mass of eggs fastened to it. On breaking into this mass,
which is brittle, it will be found to include from three hundred to five hundred eggs, about one-twenty-fifth of an inch
in diameter, of a white color, nearly globular, and flattened
on the upper side. They are placed in three or four layers,
the interstices being filled with a frothy, gelatinous matter,
which makes them adhere securely together, and over all is
a thick coating of the same material, with a nearly smooth
grayish-white surface, of a convex form, which effectually

prevents the lodgment of any water on it. The egg-mass is attached to an empty gray cocoon, the former abode of the female which deposited them.

About the middle of May the eggs hatch, when the young caterpillars at once proceed to devour the leaves of the tree on which they are placed, when disturbed letting themselves down by a silken thread, remaining suspended until danger is past, when they climb up the thread and regain their former position. When mature, they are very handsome, and present the appearance shown in Fig. 50, are more than

Fig. 50.

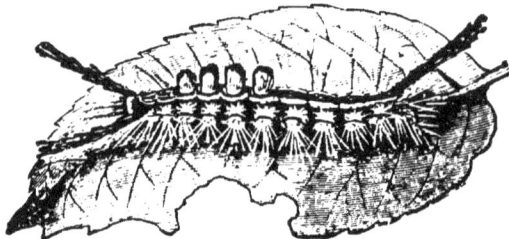

un inch long, of a bright-yellow color, with the head and two small protuberances on the hinder part of the back of a brilliant coral-red. Along the back there are four cream-colored brush-like tufts, two long black plumes on the anterior part of the body, and one on the posterior. The sides are clothed with long, fine yellow hairs. There is a narrow black or brown stripe along the back, and a wider dusky stripe on each side. There are two broods during the season, the first completing their larval growth and spinning their cocoons about the middle of July; the second hatching towards the last of July and completing their growth by the end of August, the moths from these latter depositing the eggs, which remain on the trees during the winter.

The cocoon, as already stated, is spun in the leaf; it is of a loose texture, gray in color, and has woven into it numerous hairs derived from the body of the caterpillar. The enclosed chrysalis is of an oval form and brown color, sometimes whitish on the under side, and is covered with short hairs or down. In about a fortnight the moth of the summer brood is hatched, when one might reasonably expect that from so handsome a caterpillar there would appear a moth with some corresponding beauty, but any such expectation is doomed to disappointment. In Fig. 51, *c* shows the chrysalis of the female, and *d* that of the male.

The female moth is wingless, or provided with the merest rudiments of wings; her body is of a light-gray color, of an

FIG. 51.   FIG. 52.   FIG. 53.

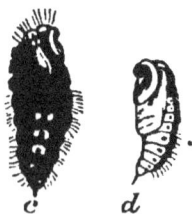

oblong-oval form, with rather long legs, and is distended with eggs; indeed, she is more like an animated bag of eggs than anything else. (See Fig. 52, where she is represented attached to the empty cocoon from which she has escaped.) After her escape, she patiently waits the attendance of the male, and then begins to place her eggs on the outside of her own cocoon, fastening them there in the manner already described. During this process her body contracts very much, and soon after her work is finished she drops down to the ground and dies.

The male moth (Fig. 53) is of an ashen-gray color, the fore wings being crossed by wavy bands of a darker shade; there is a small black spot on the outer edge near the tip, an oblique blackish stripe beyond it, and a minute white crescent near the outer hind angle. The body is gray, with a small black tuft near the base of the abdomen. The wings, when expanded, measure about an inch and a quarter across.

Since the female is wingless, and invariably attaches her eggs to the outside of her own cocoon, the insect can only spread by the wanderings of the caterpillars, or the careless introduction of eggs on young trees. No doubt the latter has been the most prolific source of mischief. Although not usually very injurious, it becomes at times a perfect pest to the fruit-grower, stripping the trees almost bare of leaves and disfiguring the fruit by gnawing its surface. While very partial to the apple, it attacks also the plum and pear, and is said to feed occasionally on the elm, maple, horse-chestnut, and oak.

Remedies.—The increase of this insect may be easily prevented by collecting and destroying the eggs during the winter months. In gathering the cocoons, all those having no egg-masses attached should be left, as they contain either the empty chrysalids of the male or the chrysalids of parasites. Nine different species of flies, four-winged and two-winged, are known to be parasitic on this insect in the caterpillar state.

No. 23.—The Yellow-necked Apple-tree Caterpillar
Datana ministra (Drury).

The moth of this species was first described by Mr. Drury, an eminent English entomologist, in 1773, from specimens received by him from New York. It measures, when its wings are expanded, about two inches across (see Fig. 54), and is of a light-brown color, with the head and a large spot on the thorax chestnut-brown.

Fig. 54.

On the fore wings there are from three to five transverse brown lines, one or two spots near the middle (sometimes wanting), and the outer margin also of the same color. The hind wings are pale yellow, without markings. When in repose, the hinder part of its body is raised up, and the fore legs stretched out. The moths appear from the middle of June until the end of July.

Each female deposits her stock of eggs in a single cluster of from seventy to one hundred in number. They are white, round, less than one-thirtieth of an inch in diameter, placed side by side in nearly straight rows, and firmly cemented to each other, as well as to the surface of the leaf on which they are placed. Those first laid begin to hatch during the third week in July, while others are three or four weeks later, so that some broods are nearly full grown, while others are small and but a few days old.

The young larvæ eat only the under side and pulpy part of the leaves, leaving the veins and upper side untouched, but as they increase in size and strength they devour the whole of the leaf except the stem. When young they are brown, striped with white, but as they mature they become darker in color, with yellow stripes ; they attain their full growth in about five or six weeks, when they are about two inches long. The head is large and black, the next segment, sometimes called the neck, of a dull orange color, a black stripe extending down the back, and three stripes of the same color alternating with four yellow stripes on each side. The body is thinly clothed with long, soft, whitish hairs. The larvæ are invariably found clustered closely together on a limb, on which, beginning with the tender leaves at the extremity, they gradually devour all before them, leaving the branch perfectly bare. Its leafless condition soon attracts attention, and on examination it is found to be loaded with these caterpillars crowded together. The position they assume when at rest is very odd, and is well shown in Fig. 55 ; both extremities are raised, the body being bent, and resting only on the four middle pairs of legs. If touched or alarmed, they throw up their heads and tails with a jerk, at the same time bending the body until the two extremities almost meet over the back ; they also jerk their heads from side to side. They all eat together, crowded upon the under surface of

the leaves, along the margins of which appears a row of shining black heads, with each mouth busily engaged in devouring the portion near it, and when
the meal is finished they arrange themselves side by side along the branches which they have stripped. If one branch does not afford food enough, they attack another; and when full grown and ready to trans-form, they nearly all leave the tree at the same time, descending by night to the ground, where they burrow under the surface to the depth of from two to four inches, and after a time cast their caterpillar skins and become naked, brown chrysalids. They remain in the pupa state until the following July, when the moths escape and take wing.

FIG. 55.

Although sometimes very abundant and destructive, this insect is not usually very common; some years a few clusters may be seen, and then several seasons may pass before they are met with again. The nakedness of the limbs they attack soon attracts attention, when the caterpillars may be easily destroyed by crushing them on the tree, or by cutting off the branches and throwing them into the fire. A small Ichneu-mon parasite is known to prey on them, which may in some measure account for the irregularity of their appearance.

No. 24.—The Red-humped Apple-tree Caterpillar.

Œdemasia concinna (Sm. & Abb.).

This insect very much resembles in habits the yellow-necked apple-tree caterpillar (No. 23).

The moth (Fig. 56) appears about the last of June. The fore wings are dark brown on the inner, and grayish on the outer margin, with a dot near the middle, a spot near each angle, and several longitudinal streaks along the hind margin, all dark brown. The hind wings of the male are brownish, or dirty white, those of the female dusky brown; the body is light brown, the thorax of a darker shade. When expanded, the wings measure from an inch to an inch and a quarter across.

The female deposits her eggs in a cluster, on the under side of a leaf, during the month of July, where they shortly hatch into tiny caterpillars, which at first consume only the substance of the under side of the leaf, leaving the upper surface unbroken, but as they increase in size they eat the entire leaf. When not eating, they remain close together, sometimes completely covering the branch they rest upon. Having come to maturity, which occurs during August or early in September, the caterpillar appears as represented in Fig. 57. The head

Fig. 56.

Fig. 57.

is coral-red, and there is a hump on the back on the fourth ring or segment of the same color; the body is traced lengthwise by slender black, yellow, and white lines, and has two rows of black prickles along the back, and other shorter ones upon the sides, from each of which there arises a fine hair. The hinder segments taper a little, and are always elevated, as shown in the figure, when the insect is not crawling. It measures, when full grown, about an inch and a quarter long.

These caterpillars entirely consume the leaves of the branch on which they are placed, and when these are insufficient the adjoining branches are laid under tribute. When handled, they discharge a transparent fluid having a strong acid smell, which doubtless serves as a defence against enemies, especially birds, since their habit of feeding openly in large flocks renders them particularly liable to attack from these ever-active foes.

When full grown, they all disappear about the same time, descending from the trees to the ground, where they conceal themselves under leaves, upon or slightly under the surface, and after a long time change to brown chrysalids, as shown in Fig. 58, and remain in the pupa state until late in June or early in July of the following year, when the perfect moths appear.

Fig. 58.

In the North there is only one brood during the year, but in the South they are said to be double-brooded. They are very generally distributed, but seldom abundant, and, while preferring the apple, feed also on the plum, cherry, rose, thorn, and pear.

As they maintain their gregarious habits during their entire larval existence, they can easily be gathered and destroyed, either by cutting off the limb and burning it, or by dislodging them by suddenly jarring the limb, when they fall to the ground and may be trampled under foot. These larvæ are also destroyed by parasites belonging to the family of Ichneumons, but it is not yet known to what species we are indebted for this friendly help.

Nos. 25 and 26.—Canker-worms.

Anisopteryx vernata (Peck), and *A. pometaria* Harris.

These are two distinct species of insects which have been confounded under the common name of canker-worm, and, as their habits and appearance are so similar, it will be convenient to treat of them under one heading. The moths from the species *pometaria* leave the ground chiefly in the fall, those of *vernata* partly in the fall, but more abundantly in the spring.

A. pometaria, known as the Fall Canker-worm, will first claim our attention. Late in the season, when many of the leaves have fallen, and severe frosts have cut everything that is tender, a walk in the woods or through the orchard on a sunny afternoon is not void of interest. Here and there slender, delicate, silky-winged moths may be seen flitting about, enjoying the sunshine. On capturing one and examining it closely, we find it to be almost transparent, and one is led to wonder why so frail a creature should select so bleak a season in which to appear; but, delicate as its structure seems to be, it is nevertheless one of the hardiest of its race, requiring, indeed, a considerable degree of cold for its perfect development. These are the male moths of the canker-worm, and chiefly those of *pometaria,* the fall canker-worm. The females are wingless.

The eggs of this species (*a* and *b*, Fig. 59) are flattened above, have a central puncture and a brown circle near the border, are laid side by side in regular masses (*e*, Fig. 59), often as many as a hundred together, and generally placed in exposed situations on the twigs or branches of the tree. They usually hatch about the time when the young leaves of the apple push from the bud, when the little canker-

Fig. 59.

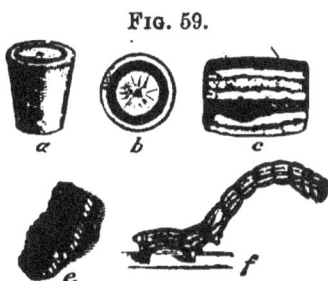

worms cluster upon and consume the tender leaves, and, on the approach of cold or wet weather, creep for shelter into the bosom of the expanding bud or into the opening flowers. The newly-hatched caterpillar is of a pale olive-green color, with the head and horny part of the second segment of a very pale hue. When full grown, it measures about an inch in length, presenting the appearance shown at *f*, Fig. 59 ; in the same figure, *c* represents a side view of one of the segments of the body, enlarged so as to show its markings. These caterpillars are called loopers, because they alternately loop and extend their bodies when in motion. They are also known as measuring-worms. They vary in color from greenish yellow to dusky or even dark brown, with broad longitudinal yellowish or paler stripes along each side. When not eating, they usually assume a stiff posture, either flat and parallel with the twigs on which they rest, or at an angle of about forty-five degrees ; in either case, since they closely resemble in color the branch on which they rest, they usually elude detection. When full grown, they leave the trees either by creeping down the trunk or by letting themselves down by silken threads from the branches. When thus suspended in great numbers, as is frequently the case, under the limbs of trees overhanging roads and sidewalks, they become a great annoyance, especially to sensitive people, and are often swept off by passing vehicles, and in this manner sometimes distributed over a considerable area.

Having reached the ground, they burrow into it to a depth of from two to six inches, where they make a rather tough cocoon of buff-colored silk, interwoven with particles of earth. The chrysalis is about half an inch long, of a light grayish-brown color, that of the male slender and furnished with wing-cases, that of the female larger and without wing-cases. The chrysalids remain in the ground throughout the summer, and the moths usually appear on the wing during the mild weather which succeeds the first severe frosts in autumn.

The female moth of each species is without wings, and sluggish in movement, with a very odd spider-like appearance.

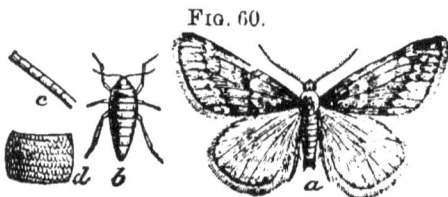

Fig. 60.

(See *b*, Fig. 60.) With a body distended with eggs, she drags her weary way along in a most ungainly manner until she reaches the base of a suitable tree, up which she climbs, and there awaits the arrival of the male. Her body is of a uniform shining ash color above, and gray beneath; it is from three to four tenths of an inch in length.

The fore wings of the male (Fig. 60, *a*) are of a brownish-gray color, very glossy, and are crossed by two rather irregular whitish bands, the outer one enlarging near the apex, where it forms a large pale spot. The hind wings are grayish brown, with a faint central blackish dot and a more or less distinct whitish band crossing them.

Anisopteryx vernata, known as the Spring Canker-worm, has an oval-shaped egg, shown at *b* in Fig. 61, highly mag-

nified; the natural size is shown in the small cluster adjoining; they are of a very delicate texture and pearly lustre, and are laid in masses without any regularity or order in their arrangement, often as many as a hundred together, usually hidden in crevices of the bark of trees. They hatch at the same time as the other species.

Fig. 61.

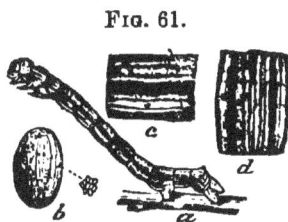

The young caterpillar is of a dark olive-green or brown color, with a black shining head, and a horny plate of the same color on the top of the next segment; they, too, are about an inch long when full grown, and present then the appearance shown at *a*, Fig. 61. In the same figure, *c* represents a side view, and *d* a back view, of one of the segments, enlarged so as to show their markings more distinctly.

When full grown, this caterpillar closely resembles that of the other species, and the body is equally variable in color. In this the head is mottled and spotted, and has two pale transverse lines in front; the body is longitudinally striped with many narrow pale lines; along the sides it becomes deeper in color, and down the middle of the back are some blackish spots. Their habits are similar to those of the other species, and they attain full growth about the same time.

The chrysalids, which are found about the same depth under ground, are similar in color to those of *pometaria,* but the cocoon is much more fragile, and is easily torn to pieces. Sometimes the moth escapes from the chrysalis in the autumn, but more frequently during the first warm days of spring.

The abdomen of the female (*b*, Fig. 62), as well as that of the male, has in this species, upon the hinder margin of each of the rings, two transverse rows of stiff reddish spines; at *d* in the figure is represented a joint of the abdomen, enlarged, showing these spines. The female also has a retractile ovipositor, shown in the figure at *e;* this is wanting in the other species ; *c* represents a portion of one of her antennæ.

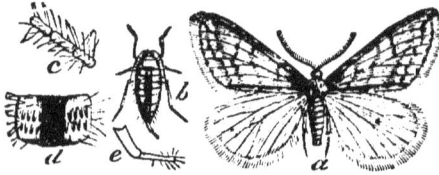

FIG. 62.

The fore wings of the male are paler than in *pometaria,* and more transparent; they are ash-colored or brownish gray, and of a silky appearance. A broken whitish band crosses the wings near the outer margin, and three interrupted brownish lines between that and the base ; there is an oblique black dash near the tip of the fore wings, and a nearly continuous black line at the base of the fringe. The hind wings are plain pale ash color, or very light gray, with a dusky dot about the middle.

Remedies.—To attack an enemy with success it is essential that we know his vulnerable points. In this instance, since the females are without wings, if they can be prevented from crawling up the trees to deposit their eggs, a great point will be gained. Various measures have been employed to secure this end, all belonging to one or other of two classes,—first, those that prevent the ascension of the moth by entangling her feet and holding her there, or by drowning her ; second, those which look to a similar end by preventing her from getting a foothold, and causing her to fall repeatedly to the ground until she becomes exhausted and dies. In the first class is included tar, mixed with oil to prevent its drying, and applied either directly around the body of the tree, or on strips of old canvas or stiff paper, about five or six inches wide, and tied in the middle with a string ; refuse sorghum molasses, printer's ink, and slow-drying varnishes, are used in a similar manner. Tin, lead, and rubber troughs, to contain oil, also belong to this class of remedies, and have all been used with more or less success. In the use of any of the first-named sticky substances, it should be borne in mind that they must be kept sticky by frequent renewal of the surface in mild weather, or the application will be useless; they should also be applied as early as the latter part of October, and kept on until the leaves are expanded in the following spring. It must also be remembered that some of the moths, defeated in their attempts to climb the trees, will deposit their eggs near the ground, or anywhere, in fact, below the barrier, and that the tiny young worms hatched from them will pass without difficulty through a very small opening. Hence, whether troughs or bandages are used, care must be taken to fill up all the irregularities of surface in the bark of the trees, so that no openings shall be left through which they may pass. Cotton batting answers well in most cases for this purpose.

The second class of remedies consists of various ingenious devices, in the way of collars of metal, wood, or glass fastened around the tree and sloping downward like an inverted funnel. These, although they prevent the moths from ascending the tree, offer but little obstacle to the progress of the young caterpillars unless the openings between the collar and the tree are carefully packed, and hence they often fail of entire success. Those belonging to the first class are said to be the surest and best, and while it must be admitted that it involves much time and labor to renew so often and for so long a period the tar or other sticky application so as to make it an effectual barrier to the ascent of the insect, still it will pay, wherever the canker-worm abounds, to give this matter the attention requisite to insure success. The limited power of motion possessed by the female usually confines this insect within narrow limits, and hence it is local in its attacks, sometimes abounding in one orchard and being scarcely known in a neighboring one; but when it has obtained a footing, and is neglected, it usually multiplies prodigiously. Strong winds will sometimes carry the larvæ from one tree to another near by. When the caterpillars are once on the tree, if the tree is small, they may be dislodged by jarring, when they all drop, suspended in mid-air by silken threads; then, by swinging a stick above them, the threads may be collected and the larvæ brought to the ground and destroyed. Fall ploughing has been recommended to destroy the chrysalids by turning them up, when they are likely to be either killed by exposure or devoured by birds. Hogs also are very useful in destroying this pest by rooting up the chrysalids and eating them.

These insects have many natural enemies. A small mite, *Nothrus ovivorus* Packard (Fig. 63), destroys the eggs. A

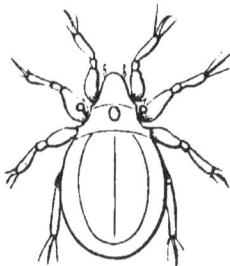

Fig. 63.

minute parasitic fly deposits her eggs within the eggs of the canker-worm and destroys them. In the larval state they are preyed on by a small four-winged fly, a species of Microgaster, which, after having fed upon its victim to full growth, eats its way out, and constructs a small oval white cocoon attached to the body of the caterpillar. A species of Tachina, a two-winged fly similar to Fig. 46, No. 21, is also a parasite on these worms. Predaceous insects also feed upon them, especially the Green Caterpillar-hunter (Fig. 47), the Copper-spotted Calosoma (Fig. 48), and the Rapacious Soldier-bug, *Sinea diadema* (Say) (Fig. 64). The

Fig. 65.

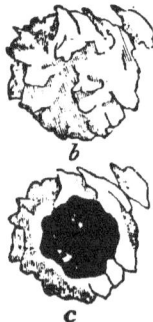

Fig. 64.

b

c

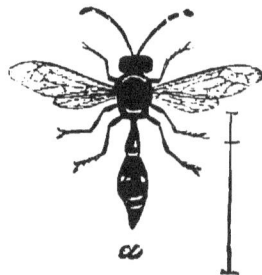

a

Fraternal Potter-wasp, *Eumenes fraternus* Say (*a*, Fig. 65), stores the cells for her young with canker-worms, often placing as many as fifteen or twenty in a single cell. In the figure, at *b* is shown the clay cell of this insect entire; at *c* the same cut through, showing how it is packed with these larvæ. These cells are sometimes attached to plants and sometimes constructed under the loose bark of trees. Insect-eating birds also devour large numbers of canker-worms.

These insects are not confined to the apple-tree: elm-trees are frequently eaten bare by them; they attack also the plum, cherry, linden, and many other trees. They are common in the Eastern and Western States, and also in some parts of Canada.

No. 27.—The Fall Web-worm.
Hyphantria textor Harris.

After the webs of the tent-caterpillars have been carefully removed in the spring, and the fruit-grower is perhaps flattering himself with the idea that his troubles in this direction are about over, towards the end of summer he may be mortified to find his trees again adorned with webs enclosing swarms of hungry caterpillars, devouring the foliage. This is the fall web-worm, an insect totally different in all its stages from the common tent-caterpillar. The moth of this species deposits her eggs in broad patches on the under side of the leaves, near the end of a branch, during the latter part of May or early in June. These hatch in the month of June, July, or August; during the earlier period in the warmer districts, and later in the colder ones.

As soon as the young larvæ appear they begin to eat, and to spin a web over themselves for protection. They devour only the pulpy portion of the leaves, leaving the veins and skin of the under surface untouched. While young, they are of a pale-yellowish color, sparingly hairy, with two rows of black marks along the body. When full grown, they are an inch or more in length, and vary greatly in their markings; some examples are pale yellow or greenish, others much darker and of a bluish-black hue. The head is black, and there is a broad dusky or blackish stripe down the back; along each side is a yellowish band, speckled more or less with black. The body is covered with long straight hairs, grouped in tufts, arising from small black or orange-yellow protuberances, of which there are a number on each segment. The hairs are sometimes of a dirty white, with a few black ones interspersed, sometimes reddish brown ; they are longest towards the extremities of the body. Unlike the common tent-caterpillars, these larvæ do not wander from their nests to feed until nearly full grown, but extend the web over their whole feeding-ground, constantly enclosing fresh portions of the branch occupied, until sometimes the web covers a space several feet long, the whole enclosed portion having a scorched or withered look, as if it had been blighted. When nearly at their full growth, they suddenly abandon their social habits and scatter far and wide, feeding on almost any green thing they meet with. They are very active, and run briskly when disturbed.

Fig. 66.

During September and October these caterpillars descend to the ground and burrow a short distance under the surface, or creep under crevices of bark or some such shelter above ground, where they form slight cocoons of silk, interwoven with hairs from their bodies. Within these cocoons they soon change to chrysalids of a dark-brown color (Fig. 67),

Fig. 67.   Fig. 68.

b

smooth, polished, and faintly punctated, with a swelling about the middle. In this condition they remain until the following year.

The moth (Fig. 68) is of a milk-white color, without spots ;

the antennæ are gray, those of the male doubly feathered below, those of the female with two rows of minute teeth only; the front thighs are tawny yellow, the feet blackish brown. When the wings are expanded they measure about one and a quarter inches across. The moth flies only at night.

In the Northern United States and Canada there is only óne brood of this insect in the season, but in the South it is frequently double-brooded, the first brood of the larvæ appearing in June, the second in August. It is a very general feeder; besides the apple, it also eats the leaves of the plum, cherry, pear, hickory, ash, elm, willow, oak, beech, buttonwood, grape, currant, blackberry, raspberry, and clover.

From their birth, the web-spinning habits of these larvæ promptly lead to their detection, and as soon as seen they should be removed by cutting off the twig or branch and destroying it; if beyond ordinary reach, the branch may be cut off by attaching a pair of pruning-shears to a pole and pulling one handle with a string. As they remain constantly under the web for so long a period, the removal of the branch insures in most instances the destruction of the whole colony.

FIG. 69

No parasites have yet been recorded as preying on them, but many carnivorous insects devour them. The Spined Soldier-bug, *Podisus spinosus* (Dallas) (Fig. 69), attacks them, piercing their bodies with its beak and sucking them empty. This friendly insect is represented in the figure at *b*, with one pair of wings extended, the other closed; at *a*, a magnified view of the beak is given.

No. 28.—The Cecropia Emperor-moth.

Platysamia Cecropia (Linn.).

Among the many beautiful insects native to this country, there are none which excite more delight and astonishment than the Cecropia moth. Its size is enormous, measuring, when its wings are spread, from five to seven inches across,

while its beauty is such as to charm all beholders.   Fig. 70
gives a very good representation of this magnificent moth.

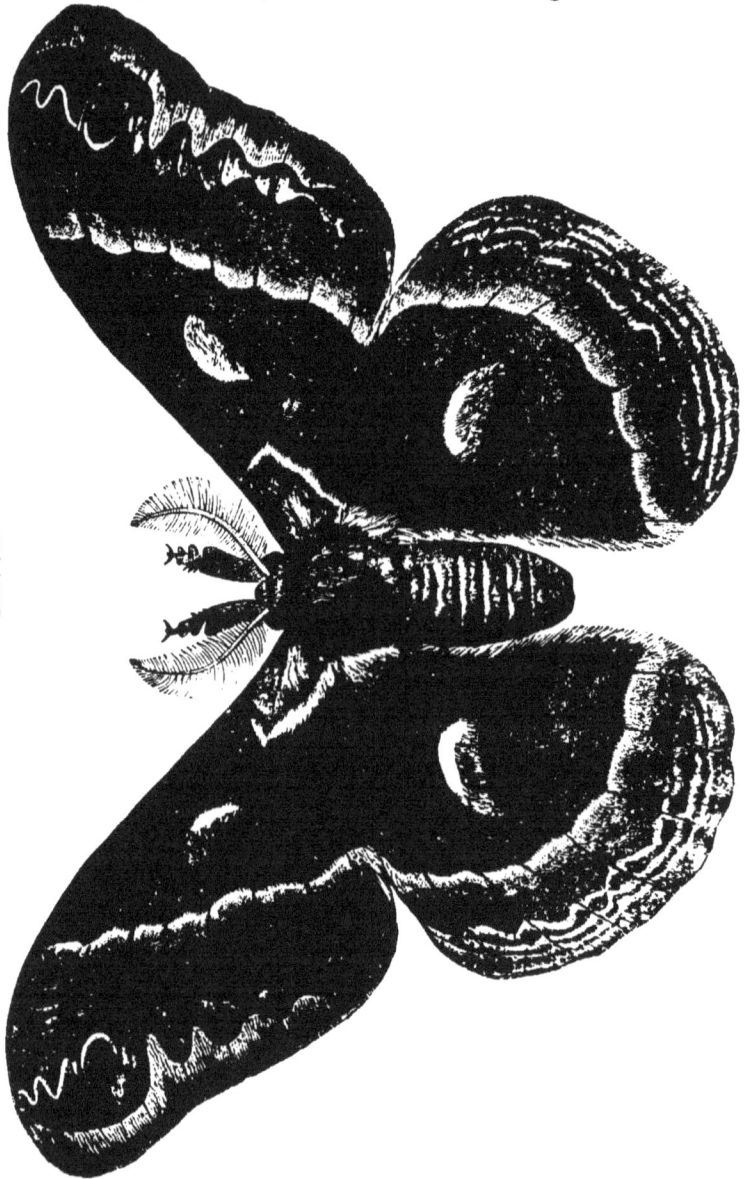

Fig. 70.

Both the front and hind wings are of a rich brown, the
anterior pair grayish shaded with red, the posterior more

uniformly brown, and about the middle of each of the wings is a nearly kidney-shaped white spot, shaded more or less with red, and margined with black. A wavy dull-red band crosses each of the wings, edged within with white, the edging wide and distinct on the hind wings, and more or less faint on the front pair. The outer edges of the wings are of a pale silky brown, in which, on the anterior pair, runs an irregular dull-black line, which on the hind wings is replaced by a double broken band of the same hue. The front wings, next to the shoulders, are dull red, with a curved white and black band, and near their tips is an eye-like spot with a bluish-white crescent. The upper side of the body and the legs are dull red, with a wide band behind the head, and the hinder edges of the rings of the abdomen white; the under side of the body is also marked with white.

During the winter months, when the apple-trees are leafless, the large cocoons of this insect are frequently found firmly attached to the twigs; they also occur on many other trees and shrubs, for in its caterpillar state it is a very general feeder. The cocoon (Fig. 71) is about three inches long and an inch or more broad in its widest part, pod-shaped, of a rusty-gray or brownish color; it is formed of two layers of silk, the outer one not unlike strong brown paper, and within this a quantity of loose silken fibres covering an inner, oval, closely-woven cocoon, containing a large brown chrysalis. Snugly enclosed within this double wrapper, the chrysalis remains uninjured by the variations of temperature during the winter. Late in May, or early in June, the pupa-case is ruptured by the struggles of its occupant, and the newly-born moth begins to work its way out of the cocoon; to lessen the labor, a fluid is secreted from about the mouth, which softens the fibres; then a tearing, scraping sound is heard, made by the insect working with the claws on its fore feet, pulling away the softened threads and packing them on each side to make a passage for its body. The place of exit is the smaller end of the cocoon, which is

more loosely made than any other part, and through which, after the internal obstacles are overcome, the passage is effected without much further trouble. First through the opening is thrust the front pair of bushy-looking legs, the sharp claws of which fasten on the outside structure; then with an effort the head is drawn forward, displaying the beautiful feather-like antennæ; next the thorax, on which are borne the other two pairs of legs, is liberated, and finally the escape is completed by the withdrawal of the abdomen. An odd-looking creature it is at first, with its large, plump, juicy body, and its thick, small wings not much larger than those of a humble-bee. The insect now seeks a good location where the wings may hang down in a position favorable for expanding, when in a short time they undergo a marvellous growth, attaining their full size in from half an hour to an hour.

FIG. 71.

Soon after their exit these moths seek their mates, and shortly the female begins to deposit her eggs, a process which occupies considerable time, since there are two or three hundred to dispose of, and they are usually laid in pairs, firmly fastened with a glutinous material, on the under side of a leaf of the tree or shrub which is to form the future food of the caterpillar. The egg is nearly one-tenth of an inch long, almost round, of a dull creamy-white color, with a reddish spot or streak near the middle.

The duration of the egg-state is usually from a week to ten days, when the young larva eats its way out, making its first meal of the empty egg-shell. At first it is black, with little shining black knobs on its body, from which arise hairs of the same color. With a ravenous appetite, its growth is very rapid, and from time to time its exterior coat or skin becomes too tight for its comfort, when it is ruptured and thrown off. At each of these changes or moultings the caterpillar appears in an altered garb, until finally it assumes the appearance represented in Fig. 72. It is a gigantic creature, from three

Fig. 72.

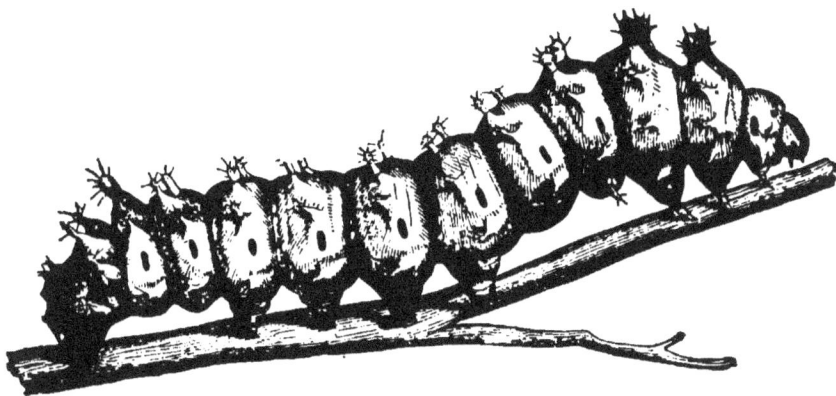

to four inches long, and nearly as thick as a man's thumb; its color is pale green; the large warts or tubercles on the third and fourth segments are coral-red, the others on the back are yellow, except those on the second and terminal segments, which, in common with the smaller tubercles along the side, are blue. During its growth from the diminutive creature as it escapes from the egg to the monstrous-looking full-grown specimen, it consumes an immense amount of vegetable food; and especially as it approaches maturity is this voracious appetite apparent. Where one or two have been placed on a young apple-tree, they may in a short time strip it entirely bare; the loss of foliage during the growing period prevents the proper ripening of the wood, and often endangers the life of the tree.

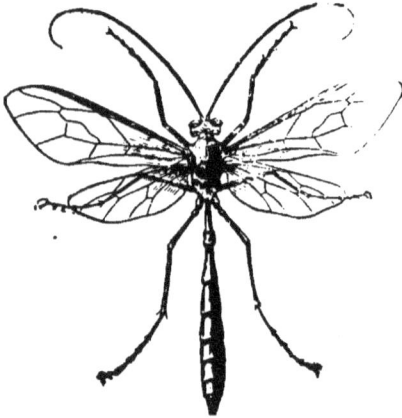

Remedies.—The natural increase of this insect is great, and wise provisions have been made to keep it within due bounds. Being so conspicuous an object, it often forms a dainty meal for the larger insectivorous birds; there are also enemies which attack the egg and young larva, and several species of parasites which live within or on the body of the caterpillar, and finally destroy it either in the larval or the chrysalis state: it is believed that fully four-fifths of the larvæ perish in this manner. The largest of these parasites, and perhaps the commonest of them all, is the Longtailed Ophion, *Ophion ma-crurum* (Linn.) (Fig. 73), a large, yellowish-brown Ichneumon.

Fig. 78.

The female of this fly deposits her eggs on the skin of her victim, where the young larvæ soon hatch, and, having firmly attached themselves, feed externally, sucking the juices of the caterpillar. After the latter has attained full growth, formed its cocoon, and become a chrysalis, this useful parasite causes its death. When full grown, the larva of the parasite is a large, fat, footless grub (Fig. 74), which spins an oblong-oval cocoon within the Crecopia chrysalis, and escapes as a fly, sometimes in the autumn, but more frequently in the following spring. A two-winged fly, a species of Tachina (Fig. 46), is also very frequently found as a parasite on the caterpillar. The larva

Fig. 74.

of this parasite is a fat, fleshy, footless grub, of a translucent yellow color, and about half an inch in length. A third parasite is a small four-winged fly, known as the Cecropia Chalcisfly, *Smicra mariæ* (Riley) (Fig. 75). In the figure the fly is

Fig. 75.

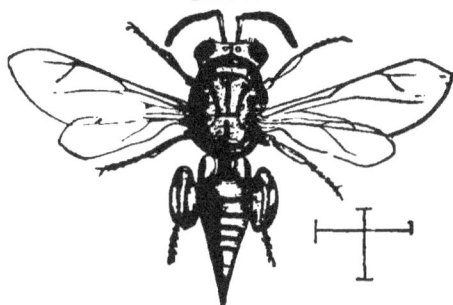

Fig 76.

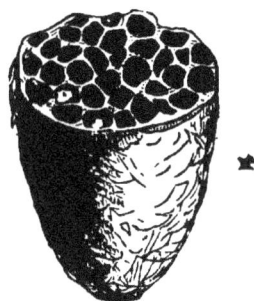

much magnified; the short lines at the side show its natural size. A fourth friendly helper is an Ichneumon fly, known under the name of the Cecropia Cryptus, *Cryptus extrematis* Cresson, which infests the Cecropia larva in great numbers, filling its chrysalis so entirely with its thin, papery. cocoons that a transverse section bears a strong resemblance to a piece of honey-comb. (See Fig. 76.) The flies of this parasite escape in June, the female presenting the appearance shown in Fig. 77, where it is much mag-

Fig. 77.

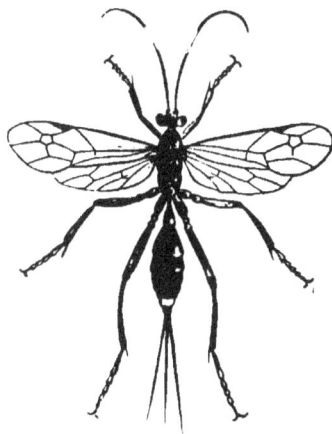

nified, the short line at the side showing its natural size. Another two-winged parasite is *Gaurax anchora* Loew.

While very partial to the apple, the larva of Cecropia will also feed on the cherry, plum, pear, maple, willow, lilac, English alder, red currant, and hazel; also on the hickory, birch, elm, honey-locust, barberry, hawthorn, and elder.

During the winter their cocoons should be looked for and destroyed; the larvæ also may be subdued by hand-picking,— their work, as well as their appearance, being so conspicuous that they are readily detected.

No. 29.—The Unicorn Prominent.

Cœlodasys unicornis (Sm. & Abb.).

The larva of this moth is a very singular-looking creature. (See Fig. 78.) It is reddish brown, variegated with white, on

Fig 78.

the back, with a large brown head; the sides of the second and third segments are green, and from the top of the fourth a prominent horn is projected. There are on the body a few short hairs, scarcely visible to the naked eye; the posterior segment, with the hindermost pair of feet, is always raised when the insect is at rest, but it generally uses these feet in walking. In August and September this larva may be found nearly full grown. At first eating a notch, about the size of its body, in the side of the leaf on which it is feeding, and placing itself in this notch, with the humps on its body somewhat resembling the irregularities in the margin of the partly-eaten leaf, it is not easily detected. Eventually it consumes the entire leaf, except a small portion of the base. When mature, it measures from an inch to an inch and a quarter in length, and, while generally solitary in its habits, sometimes three or four are found together eating the leaves of the same twig. Besides the apple, it feeds on the plum, dogwood, rose, alder, and winterberry.

When full grown, which is towards the end of September, it descends from the tree, and under fallen leaves on the ground constructs a thin, almost transparent, papery cocoon, with bits of leaves attached to the outside. A considerable time elapses after the cocoon is formed before the caterpillar changes to a brown chrysalis. The moth does not appear until the following summer, and is most common in July. (See Fig. 79.)

The fore wings are light brown, variegated with patches of greenish white, with many wavy lines of a dark-brown color, two of which enclose a small whitish space; at the base there is a short blackish mark near the middle; the tip and the outer hind margin are whitish, tinged with red in the males, and near the outer hind angle there are two black dashes and one small white dash. The hind wings of the male are dirty white, with a dusky spot on the inner hind angle, those of the female sometimes entirely dusky. The body is brownish, with two narrow black bands across the front part of the thorax. When the wings are expanded, this moth measures from an inch and a quarter to an inch and a half across. It is double-brooded in the South, the moths of the first brood appearing early in June, those of the second in August; in the North it is also sometimes double-brooded.

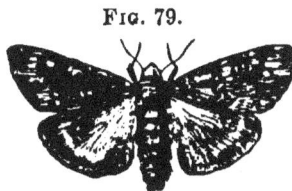

Fig. 79.

This insect is rarely present in sufficient numbers to do any material damage; and it seldom attracts the notice of the fruit-grower, unless by the singular appearance of the caterpillar and its remarkable combination of colors. No parasites have yet been recorded as preying on it, though doubtless it suffers in this way in common with most other insects.

No. 30.—The Turnus Swallow-tail.

Papilio turnus Linn.

Every one must have seen the large turnus swallow-tail butterfly floating about in the warm days of June and July, enjoying the sunshine, drinking from the wayside pool, or sipping the honey from flowers. It is one of our largest and handsomest butterflies, measuring, when its wings are ex-

panded, about four inches across. (See Fig. 80.) The wings
are of a rich, pale lemon-yellow color, banded and bordered
with black; on the fore wings are four black bars, the inner
one extending entirely across the wing, the outer ones becoming shorter as they approach the apex. The front margin is edged with black, and the outer margin has a wide
border of the same, in which is set a row of eight or nine
pale-yellow spots, the lower ones less distinct.

FIG. 80.

The hind wings are crossed by a streak of black, which is
almost a continuation of the inner band on the fore wings;
there is a short black streak a little beyond, and a wide black
border, widening as it approaches the inner angle of the
wing. Enclosed within this border, and towards its outer
edge, are six lunular spots, the upper and lower ones reddish,
the others yellow; above and about these spots, and especially
towards the inner angle of the wing, the black bordering is
thickly powdered with blue scales. The outer margin of the
hind wings is scalloped and partly edged with yellow; the
inner margin is bordered with brownish black for about two-

thirds of its length, followed by a small yellow patch, which is succeeded by a larger black spot, centred with a crescent of blue atoms, and bounded below by an irregular reddish spot, margined within with yellow. The hind wings terminate in two long black tails edged on the inside with yellow. The body is black above, margined with pale yellowish; below, yellowish streaked with black. The under surface of the wings resembles the upper, but is paler.

This insect passes the winter in the chrysalis state, and appears first on the wing from the middle to the end of May, but becomes more plentiful during the latter part of June and early in July. The eggs are deposited singly on the leaves of the apple and other trees and shrubs on which the larva feeds; they are about one-twenty-fourth of an inch in diameter, nearly round, of a dark-green color, with a smooth surface. In about ten or twelve days the eggs begin to change color, becoming darker, and growing very dark just before the escape of the larvæ. The very young caterpillars are black, roughened with small brownish-black tubercles, with the first segment thickened, of a dull, glossy flesh color, a prominent fleshy tubercle on each side, and a patch of white on the seventh and eighth segments.

When full grown, it appears as in Fig. 81. It is then from an inch and a half to two inches long, with a rather large reddish-brown head, and a green body, which is thickest towards the head and tapers posteriorly. On the anterior segments the

Fig. 81.

green is of a darker shade, but paler on the sides of the body, and partly covered with a whitish bloom. On the front edge of the first segment is a raised yellow fold, which slightly overhangs the head, and from which, when irritated, the larva protrudes a yellow, fleshy, forked organ, at the same time giving off a disagreeable odor, which is doubt-less used as a means of defence against its enemies. On each side of the third segment is an eye-like spot, nearly oval, yellow, enclosed by a ring of black, centred with a small elongated blue dot, which is also set in black. On the hinder portion of the fourth segment is another raised yellow fold, bordered behind with rich velvety black; the latter is seen only when the larva is in motion. On the terminal segment there is a similar fold, flattened above, with a slight protu-berance on each side. On the fifth segment are two blue dots, one on each side, and there are traces on the hinder segments of similar dots, arranged in longitudinal rows. The under surface is paler than the upper, with a whitish bloom.

When the caterpillar is about to change to a chrysalis, which is usually during the early part of August, the color of the body grows gradually darker, until it becomes dark reddish brown, with the sides nearly black, and the blue dots become much more distinct. Having selected a suitable spot in which to pass the chrysalis state, it spins a web of silk, into which the hooks on the hind legs are firmly fastened; then, having prepared and stretched across a silken band or loop to support its body in the middle, it casts its larval skin, and remains a dull-brown chrysalis, of the form shown in Fig. 82, until the following spring.

FIG. 82.

This insect is very widely distributed, being found in nearly all parts of the United States and Canada. The caterpil-lar feeds on a number of different trees, but chiefly affects the apple, cherry, thorn, and basswood. As it is always solitary in its habits, it is never likely to cause much injury. South of Pennsylvania the female of this species of butterfly usually loses its yellow color and becomes nearly black, while the other sex retains its normal hue.

No. 31.—The Blind-eyed Sphinx.

Smerinthus excæcatus (Sm. & Abb.).

During September, and sometimes as late as the beginning of October, there may be found occasionally on the apple-tree, feeding on the leaves, a thick, cylindrical caterpillar, about two and a half inches long, with a green triangular head, bordered with white, an apple-green body, paler on the back, but deeper in color along the sides, with its skin roughened with numerous white-tipped granulations, having a stout horn on the hinder part of its back, of a bluish-green color, with seven oblique stripes on each side, of a pale yellow, the last one of a brighter yellow than the others and extending to the base of the horn. This is the larva of the blind-eyed sphinx, represented in Fig. 83.

Fig. 83.

When full grown, it leaves the tree and buries itself in

Fig. 84.

the earth, where it changes to a chrysalis of a chestnut-brown color, smooth, with a short terminal spine.

The moth (Fig. 84) appears from May to July, but chiefly

in June, and is very handsome. The body is fawn-colored ; on the top of the thorax is a chestnut-colored stripe, and on the abdomen a dark-brown line. The front wings are fawn-colored, clouded and striped with brown ; the hind wings are rose-colored in the middle, with a brownish patch at the tip, crossed by two or three short whitish lines, and having near the inner angle a black spot with a pale-blue centre. This moth measures, when its wings are spread, about three inches across.

It is comparatively a rare insect, and has never been known to cause any serious injury. While partial to the apple-tree, the caterpillar will also feed on the plum and wild cherry. The moth remains hidden during the day, but becomes very active at dusk.

No. 32.—The Apple Sphinx.

Sphinx Gordius Cram.

This insect belongs to the same family as No. 31, viz., the Sphingidæ, or Sphinx family, and there is a general resemblance between the two species in all their stages. The larva of the apple sphinx is a thick, cylindrical, apple-green caterpillar, about two and a half inches long, with a reddish-brown horn projecting from the hinder part of its back, and with seven oblique stripes along each side, of a violet color, margined behind with white.

Late in the autumn it leaves off feeding and buries itself deeply in the earth, where it changes to a brown chrysalis with a short detached tongue-case. Here it remains until the following season.

The perfect insect is a strong, narrow-winged moth, which appears on the wing from the latter part of May to the end of June. (Fig. 85.) Its fore wings are dark brown, varied with ash-gray, with black streaks within the veins, and a white dot near the middle, resting on a long black line. The hind wings are gray, with a band across the middle, and a wide marginal band of black. The fringes of the wings are white, the head and thorax blackish brown. The abdomen is dark gray, with a central black line, and alternate black and grayish bands partly encircling it. When the wings are ex-

FIG. 85.

panded, the moth measures from three to three and a half inches across. This also is a night-flyer.

No. 33.—The American Lappet-moth.

Gastropacha Americana Harris.

This singular insect is found in the larval state in July and August, resting in the daytime on the twigs or limbs of the apple-tree, feeding at night. Its body is broad, convex above, and perfectly flat beneath, and when at rest it closely resembles a natural swelling of the bark. It is of an ash-gray color, fringed close to the under surface on each side with tufts of blackish and gray hairs springing from projecting tubercles. On the hinder part of the third segment there is a bright-scarlet velvety band, and a similar one on the fourth segment, neither of which is seen except when the larva is in motion. On the second segment there are two small tubercles on each side, and one on each side of the remaining segments; from these tubercles are given out tufts of grayish hairs mingled with white ones. The under side of the body is orange-colored, with a central row of diamond-shaped blackish spots. In general appearance it much resembles Fig. 87.

When ready to transform, it attaches itself to a limb and there encloses itself in a gray cocoon, which appears like a slight swelling of the limb, and in this enclosure it changes to a brown chrysalis, in which state it remains until the month of June following, when the perfect insect escapes.

The moth (Fig. 86) is of a tawny reddish-brown color, with the hinder and inner edges of the fore wings and the outer edges of the hind wings notched; the notches are margined with white. Both pairs of wings are crossed by a rather broad, interrupted, whitish band, not very clearly shown in the figure, which, on the anterior wings, does not always extend to the front margin. In the female the pale bands and dark lines are sometimes wanting, the wings being almost entirely of a red-brown color. The moth measures, when its wings are expanded, from an inch and a half to an inch and three-quarters across.

FIG. 86.

The eggs are laid on the leaves of the apple-tree late in June, and are very pretty objects under a magnifying-glass. They measure about one-twentieth of an inch long, are oval, flattened at the base and also above, and a little thicker at one end than at the other. In color they are white, with peculiar black markings; at each end is a crescent-shaped stripe, with a dot below it, and on both the flattened surfaces there are markings like eyes, each formed by an oval spot in the centre, with a curved stripe above and a shorter straight one below; between and parallel to the two eyebrow-like marks there is another black stripe. The whole surface is covered with a net-work, the meshes of which are irregular, with a depressed dot in the centre of each. This insect feeds also on the cherry and the oak. It is not at all common, and probably will never be a source of much annoyance to the fruit-grower.

No. 34.—The Velleda Lappet-moth.

Tolype velleda (Stoll).

The caterpillar of this species is very similar in appearance and habits to that of the American Lappet-moth, No. 33, with some slight differences in color and markings. The full-grown larva is two inches or more in length, with a small, flat head, nearly hidden beneath two projecting tufts of hair from the second segment. It is represented partly grown in Fig. 87. The body is bluish gray, with many faint paler longitudinal lines; across the upper part of the fourth segment there is a narrow velvety black band, more conspicuous

FIG. 87.

when the caterpillar is in motion. On each segment above there are two warts with short black hairs, of which those on the fourth segment, anterior to the band, are most prominent. There are a few short black and gray hairs scattered over the body. The side fringes which border the body close to the under surface are composed of spreading tufts of light-gray mingled with black hairs, of unequal length, proceeding from warts nearly one-tenth of an inch long. The under side is of a pale-red or orange color, with black spots. This caterpillar, when at rest, closely resembles the color of the twig to which it is attached, and hence is difficult to detect. It reaches maturity during the month of July, and is found on the cherry and elm, as well as on the apple.

The cocoon, which is usually attached to one of the branches of the tree on which the larva has fed, is about an inch and a half long and half an inch wide, oval, convex above, and flattened on the under side; it is of a brownish-gray color, with a few blackish hairs interwoven with the silk.

The moth (Fig. 88) is usually found in August and September. It has a large, thick, woolly body, of a white color, variegated with bluish gray; its legs are thick and very

hairy.   On the fore wings are two broad, dark-gray bands, intervening between three narrow, wavy, white bands; the veins
are white and prominent.   The
hind wings are gray, with a white
hind border, and across the middle
there is a broad, faint, whitish
band.   On the top of the thorax
is an oblong, blackish-brown spot,
widening behind.   The males are
not much more than half the size of the females; the former,
when their wings are expanded, measure about an inch and a
half across, the latter nearly two and a half inches.   Like
that last described, this is a rare insect, and one never likely
to appear in sufficient numbers to be troublesome.

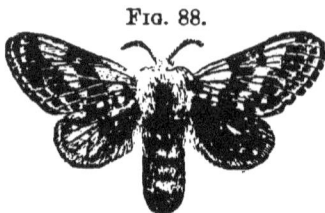

FIG. 88.

No. 35.—The Oblique-banded Leaf-roller.
Cacœcia rosaceana (Harris).

This moth is a member of a very large family of small
moths called Tortrices, or, popularly, leaf-rollers, because
their larvæ have the habit of rolling up the leaves, or portions of them, forming hollow cylinders, firmly fastened with
silken threads, in which they live, and where they are partly
protected from birds and other enemies.   Most of these
insects, when disturbed, slip quickly out of their enclosure
and let themselves down to the ground by a fine silken thread,
and thus frequently escape danger.

Soon after the buds of the apple-tree begin to open, the
caterpillars of the oblique-banded leaf-roller commence their
labors.   They coil up and fasten together the small and tender
leaves, which thus furnish them at once with shelter and food.
When full grown, they are about three-quarters of an inch
in length, of a pale-green or yellowish-green color, sometimes
reddish or brownish, with the head and top of the first segment brown; there is usually a darker green stripe along the
back, and a few smooth dots on each segment, from each of
which there arises a short, fine hair.   In Fig. 89 this larva is

shown somewhat magnified; also the chrysalis, which is about
the natural size. Besides consuming the leaves, this leaf-
roller is very fond of gnawing the
skin of the young fruit, and such
abraded spots soon become brown
and rusty, and sometimes crack.

Fig. 89.

When mature, the larva lines the
inner surface of its dwelling-place
with a web of silk, and then changes
to a chrysalis of a dark-brown color.
(See Fig. 89.) Towards the end of
June, or early in July, with the help of some little thorns
on the hinder segments, the chrysalis wriggles itself half-
way out of the nest, and shortly after the imprisoned moth
escapes.

This is a short, broad, flat moth, resembling a bell in
outline when its wings are closed (see Fig. 90); but when
expanded (Fig. 91), they appear arched on the front edge,

FIG. 90.

FIG. 91.

curving in a contrary direction near the tip. The body is
reddish brown, the fore wings of a light cinnamon-brown
color, crossed with little, wavy, darker brown lines, and with
three broad, oblique, dark-brown bands, one of which covers
the base of the wings and is sometimes indistinct or wanting; the second crosses the middle of the wings; and the
third, which is broad on the front edge and narrow behind,
is near the outer hind margin. The hind wings are ochre-
yellow, with the folded part next to the body blackish.
When the wings are expanded, the moth measures about an
inch across. The caterpillars are found on the apple, pear,

plum, peach, cherry, rose, raspberry, gooseberry, currant, strawberry, and probably some other plants, shrubs, and trees.

Remedies.—In the larval state this insect is infested by a parasite, a species of Ichneumon. A single parasite almost fills the body of the caterpillar, and yet the latter goes on actively feeding, and grows to maturity without showing any signs of inconvenience. When about to enter the chrysalis state, the occupant eats its way out of the body of its victim, which shrinks up and dies, and the parasite spins a cocoon within the leafy enclosure, and forms a chrysalis nearly as large as that of the leaf-roller, from which, in due time, a four-winged fly escapes.

The depredations of this foe are sometimes serious, more especially when it selects as its abode the terminal branches of the tree, and thus checks its growth. Whenever practicable, the curled and twisted clusters of leaves should be pinched and the larvæ crushed; if out of reach, syringing with powdered hellebore and water, in the proportion of an ounce to a pailful of water, or with Paris-green and water, in the proportion of a teaspoonful to a pailful of water will destroy many of them.

No. 36.—The Lesser Apple-leaf Folder.
Teras minuta (Robs).

The caterpillar of this species is a small greenish larva, smooth, with a pale-brown head and whitish markings. Those of the first brood make their appearance with the opening foliage in spring; the opposite edges of the tender leaves are drawn together upwards, and fastened with a silken web, thus forming a roof over the insect, which serves the double purpose of shelter and protection. The second brood, hatching later in the season from eggs laid on the surface of the mature and less yielding leaf, do not draw its edges together, but simply construct a web over the surface of the leaf. When mature, the caterpillar eats off the upper cuticle of part of a leaf, and brings the edges together, tying them with silken threads, and then lines the enclosure with fine white silk.

Within this curled leaf the caterpillar changes to a brown chrysalis, about three-tenths of an inch long. Some of the segments of the body are furnished with minute spines, and the posterior extremity with two hooks, bent downwards, with which the pupa works itself half-way out of the enclosure before the moth escapes.

The moth is about one-third of an inch long, and measures, when its wings are spread, half an inch or more across. Its head, thorax, and fore wings are of a bright-orange color, the hind wings, body, and legs whitish, with a silken lustre. The first moths appear early in the season, in time to deposit their eggs on the young foliage as it bursts the buds; the second brood appear during the latter half of July.

This insect sometimes occurs in great numbers, destroying the leaves of apple-trees, particularly young trees, giving them the appearance of being scorched by fire. When it becomes necessary to destroy them, the remedies mentioned under No. 35 should be promptly applied.

No. 37.—The Leaf-crumpler.

Phycis indigenella (Zeller).

The fruit-grower will frequently find, on examining his apple-trees in winter, clusters of curious little cases, partly hidden by portions of crumpled and withered leaves. The cases (Fig. 92, *a*, *b*) resemble long miniature horns, wide at one end, tapering almost to a point at the other, and twisted in a very odd manner. The withered leaves are firmly fastened to the cases and to the twig by silken threads, and the case itself, which is attached to the bark of the twig on which it is placed, is curiously constructed of silk interwoven with the dried castings of the artificer. The inner surface of the case is whitish and smooth, the exterior rougher and of a yellowish-brown color.

These odd little cases are the work of the larvæ of the

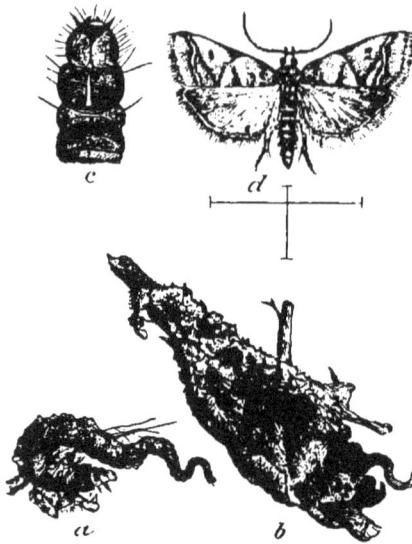

Leaf-crumpler, the young of which appear late in the summer and attain about one-third of their growth before winter sets in. After constructing their places of abode, they remain in them all winter in a torpid state. Fig. 93 represents one of these cases well covered with withered leaves. As soon as the warmth of a spring sun causes the buds to expand, the caterpillar resumes its activity, and, leaving its case in search of food,—for which purpose it usually chooses the night-time,—it draws the opening leaves towards its case, so as to secure a safe retreat should danger threaten, and, fastening them by threads of silk, enjoys its meals in comparative safety. Its length, when full grown, is about six-tenths of an inch, the body tapering slightly towards the hinder extremity. The head is dark reddish brown, and the body a dark, dull greenish brown; the first segment has a horny plate at the top, and a flattened blackish prominence on each side, below the plate; on each of the other segments there are several small blackish dots, from every one of which there arises a single brown hair. At c, Fig. 92, the head and anterior segments of this caterpillar are shown.

Fig. 92.

Fig. 93.

By the early part of June its growth is completed. It then shuts itself up in its case and changes to a reddish-brown chrysalis, about four-tenths of an inch long, from which, in about two weeks, the perfect moth escapes.

When its wings are expanded, the moth (see *d*, Fig. 92) measures about seven-tenths of an inch across. Its fore wings are pale brown, with patches and streaks of silvery white, the hind wings plain brownish white; the under side of both wings is paler. There is only one brood during the year, the moths depositing their eggs during July.

Remedies.—One would imagine that a caterpillar protected as this one is, within its case, would be secure from all enemies, but it is not so; a small Ichneumon fly is a parasite upon it; so, also, is a two-winged Tachina fly, *Tachina phycitæ* (Le Baron), which closely resembles the common house-fly.

It is not often that this insect is very numerous in any one orchard, but where it is abundant it sometimes inflicts a considerable amount of damage, consuming the young foliage and materially retarding the growth of the tree. The only way to destroy them is to pick the cases with the crumpled leaves off the trees during the winter and burn or crush them. Besides the apple, it feeds on the cherry, quince, and plum, and occasionally on the peach.

No. 38.—The Eye-spotted Bud-moth.

Tmetocera ocellana (Schiff).

The caterpillar of this insect selects the opening bud as its point of attack. It is a small, cylindrical, naked larva (see Fig. 94), about three-quarters of an inch in length, of a pale, dull, brownish color, with small warts on its body, from which arise fine short hairs; the head and the top of the next segment are black. Its tenement consists of a dried, blackened leaf, portions of which are drawn together so as to make a rude case, the central part of which is lined with silk. It is very partial to

Fig. 94.

the blossoms and newly-formed fruit, thereby causing great disappointment to fruit-growers, who have perhaps waited patiently for years for the fruit of some new or interesting variety, and have their hopes excited by seeing, it may be, a single bunch of blossoms set well and appear promising, when this mischief-maker commences its depredations on the young fruit, drawing the several portions together with threads of silk, and partly devouring them. It sometimes contents itself with injuring the leaves only, drawing one after another around its small inside case until there is formed a little cluster of withered and blackened leaves. Another of its tricks is to gnaw a hole into the top of the branch from which a bunch of blossoms issues, and, tunnelling it down the centre, cause its death.

These larvæ are usually full grown by the middle of June, when they change to dark-brown chrysalids within their nests, from which the perfect insects escape in July.

The moth (Fig. 94) measures, when its wings are expanded, about half an inch across. It is of an ash-gray color. The fore wings have a whitish-gray band across the middle, and there are two small eye-like spots on each of them, one, near the tip, composed of four little black marks on a light-brown ground, the other, near the hind angle, formed by three minute black spots arranged in a triangle, with sometimes a black dot in the centre. The hind wings are dusky brown.

The attacks of this insect are not restricted to the apple; it is injurious also to the cherry and plum. Small and insignificant as it appears, it is capable of much mischief. The only remedy suggested is to pull off and crush the withered clusters of leaves containing the caterpillars or chrysalids early in the spring. .

No. 39.—The Apple-bud Worm.

Eccopsis malana Fernald.

This insect, recently recorded as injurious, has seriously injured the apple-trees in the orchards of Northern Illinois, by devouring the terminal buds on the branches. In the larval state the mischief is done; it is then a small pale-greenish or yellowish-green caterpillar, sometimes tinged with pink on the back. Its head is yellowish, with a black dot on each side, and there is a patch or shield of a yellowish color on the upper part of the next segment.

The eggs from which these caterpillars hatch are deposited singly upon the terminal buds. The young larva, after devouring the bud, fastens the leaf-stalk of one of the leaves growing near the tip to the side of the branch, and thus forms for itself a sort of burrow between the leaf-stalk and the branch, in which it hides during the day, issuing from its retreat at night to feed on the leaf so secured. When this is consumed, it is said to feed for a time on the newly-formed wood, and sometimes eats its way a short distance into the twig. The caterpillar about this time deserts its burrow on the branch, and constructs a yellow, woolly tube or case upon one of the leaves, in which it lives, issuing at night to feed as heretofore, and when the leaf on which it is placed is almost consumed, the larva drags the case to an adjoining leaf. As it approaches maturity, it becomes of a dark flesh-color; its body is marked with a number of small shining spots, and its head and the horny shield on the next segment are black. When full grown, it measures about half an inch in length; it then closes its case with a silken lid and changes to a chrysalis within it, from which the moth appears about a week or ten days later.

The fore wings of the moth are white, mottled and spotted with greenish brown; there is a large grayish-brown spot at the tip, mottled with white, and another, towards the base of the wing, of a darker shade; the front edge is mottled with grayish brown. The hind wings are dusky. There is only one brood of these insects during the year.

The tips of the infested branches usually die back as far as the base of the first perfect leaf, where a new bud forms, which takes the place of the terminal bud. As the branch

7

from this new-formed bud is late in starting, and does not grow straight, the injury caused by this insect interferes seriously with the growth of the tree, and also mars its beauty.

A small Ichneumon fly, *Microdus earinoides* Cresson, attacks this bud-worm, depositing an egg in the body of each caterpillar, which, hatching, produces a footless larva, that lives within the body of the caterpillar until it is about ready to become a chrysalis, when the larva issues from its body and the caterpillar dies. The parasite spins within the silken case of its host a tough white cocoon about one-fourth of an inch long, from which the perfect fly issues in about a fortnight.

Where these insects are very troublesome they may be destroyed by syringing the trees with Paris-green or London-purple mixed with water, in the proportion of one or two teaspoonfuls of the poison to two gallons of water. Their numbers may also be lessened by hand-picking, gathering them while still in their burrows near the tops of the twigs.

No. 40.—The Green Apple-leaf-tyer.

Teras minuta (Robs) *var. Cinderella* (Riley).

This is a small yellowish-green caterpillar (*a*, Fig. 95), with a horny head and neck of a deeper yellowish shade, the head being marked with a crescent-shaped black mark. It

Fig. 95.

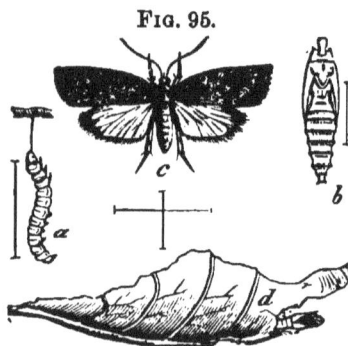

belongs also to the leaf-rollers or leaf-folders, and draws the edges of the leaf together, as shown in the figure at *d*, and lives within the fold. In feeding, it eats the leaf entirely through. It is a very nimble little creature, and when disturbed wriggles quickly out of its case and drops to the ground.

The larva changes to a brown chrysalis (*b*, Fig. 95) within the fold of the leaf, which is lined with silk. When the time approaches for the moth to escape, the chrysalis wriggles itself so far out that the head projects beyond the enclosure, as shown at *d*, soon after which the moth appears.

The front wings of the moth (*c*, Fig. 95) are of a glossy, dark ash-gray color, the hind wings a little paler; when its wings are spread, it measures about an inch across.

This insect is but a slate-colored variety of No. 36, but sufficiently marked in its character to justify a description under a separate heading.

No. 41.—The Apple-leaf-sewer.

Phoxopteris nubeculana (Clem.).

In the perfect state, this insect is a small moth belonging to the Tortricidæ, or Leaf-rollers. It passes the winter in the larval condition in rolled-up apple-leaves which lie on the ground. Early in April the larvæ change to chrysalids, and about ten days afterwards the moths begin to appear, and continue to issue for several weeks.

The moth is white, with brown markings, as shown in Fig. 96, at *c*. The eggs are laid in June, and the larva is found

FIG. 96.

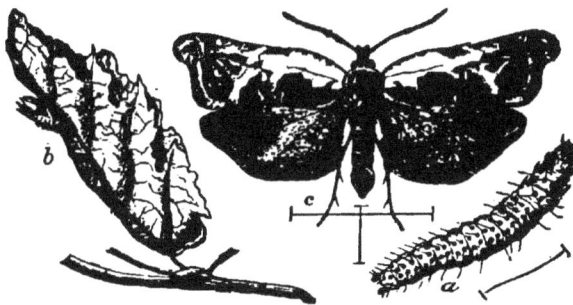

throughout the summer and autumn on apple-leaves. It folds the leaves together, as shown at *b* in the figure, making the edges meet, so that the whole leaf forms a hollow case, within which it lives and feeds on the softer tissues. The larva is of a yellowish-green color, with a yellow head, and with a horny shield on the next segment, a little darker, with a black dot on each side. On each of the remaining segments there are a number of pale, shining, raised dots, from every one of which arises a single hair. On the approach of winter the larva lines its chamber with silk, and falls with the leaf to the ground, where it remains unchanged until early the following spring, when it becomes a yellowish-brown chrysalis. As the time approaches for the escape of the moth, the chrysalis wriggles its way through the partly-decayed leaf-case at the back, and protrudes as shown at *b* in the figure, soon after which the moth escapes.

This caterpillar sometimes prevails to such an extent as seriously to injure the foliage of apple-trees; in such cases the most obvious remedy is to gather carefully in the autumn all the fallen leaves with the enclosed larvæ and burn them.

No. 42.—The Apple-leaf Skeletonizer.

Pempelia Hammondi Riley.

This insect occurs in the larval state in the autumn, and sometimes during the summer also, and is especially injurious to young orchards and nurseries, giving the foliage a rusty, blighted appearance, caused by the larva devouring the green pulpy parts of the upper surface of the leaves and leaving the closely-netted veins with the under skin untouched. The larva (Fig. 97, *a*) is of a pale-brownish color, about half an inch long, with darker lines, as shown at *b*, where one of the segments is highly magnified; sometimes the color assumes a greenish shade. Behind the

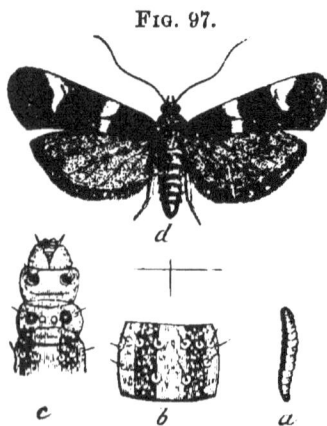

Fig. 97.

head there are four shiny-black tubercles, as shown at *c* in the figure, also magnified. The larva covers the surface of the leaf with loose silky threads, attached to which will be found a number of small black grains of excrementitious matter, and under this rough covering the larva feeds. It sometimes feeds singly and sometimes in groups; in the latter case a number of the leaves are drawn together, and the caterpillars live and feed within this shelter.

The chrysalis is usually formed among the leaves in a very slight cocoon, and is about a quarter of an inch long and of a pale-brown color. The winter is passed in the chrysalis state, and the moths appear during May or June following.

When its wings are spread, the moth measures nearly half an inch across; it is of a deep purplish-gray color, with a glossy surface, and has two silvery-gray bands across the wings, as shown in the figure, at *d*, where it is magnified; the cross-lines below the figure indicate the natural size.

Remedies.—This pest may be subdued by hand-picking if begun in good season. It is preyed on by two species of small Ichneumon flies, and by several carnivorous insects.

No. 43.—The Many-dotted Apple-worm.

Nolaphana malana (Fitch).

In June, and again in August or September, there is sometimes found on apple-leaves, in considerable numbers, a rather thick, cylindrical, light-green larva, an inch or more in length, with five white longitudinal lines and numerous whitish dots. These are the larvæ of *Nolaphana malana*. They eat irregular notches in the margins and holes in the middle of the leaves, and do not feed in groups, but are solitary in their habits, scattered among the foliage. They begin to appear about the last of May, and live openly exposed on the under side of the leaves, without forming any web or fold in the leaf for protection. On reaching maturity, which for the early brood is about the last of June, the larva selects a leaf and draws together a portion of it with silken threads, forming a hollow tube, within which it spins a slight silky cocoon and changes to a brown chrysalis. In this inactive condition the insect remains for three or four weeks, sometimes longer, when the moth appears.

The moth (Fig. 98) is a very pretty object. Its fore wings are ash-gray, whitish towards the outer margin, and crossed

Fig. 98.

by three irregular black lines, which are faint or indistinct towards the inner edge; near the middle of the wing there is often a round, whitish spot, with a black dot in the middle. The hind wings are dull-whitish, dusky towards the tips. Beneath, both wings are of a silvery-whitish hue, sprinkled with blackish dots towards the outer edges. When the wings are expanded, they measure from three-quarters of an inch to an inch or more across. .

The first moths appear early in spring, and attach their eggs to the young foliage; the second brood appear in July. These attach their eggs to the leaves, and produce larvæ in August and September, which, when their growth is completed, change to chrysalids within the folded leaves, as already described, and are carried to the ground with the fall of the leaves in autumn, where they pass the winter in the pupa state and produce moths in the following spring.

These larvæ feed also on cherry, peach, elm, poplar, and other trees. They are seldom sufficiently numerous to be troublesome, but if at any time a remedy is required they may be destroyed by syringing the leaves with Paris-green or hellebore mixed with water, as recommended for No. 35. When the trees on which they are feeding are suddenly jarred, the caterpillars will drop to the ground, and by taking advantage of this peculiarity they may be captured and destroyed.

No. 44.—The Palmer-worm.

Ypsolophus pometellus (Harris).

This larva appears on apple-trees during the latter part of June, and at times is excessively numerous and destructive.

It lives in societies, making its home in a mass of half-eaten and browned leaves, drawn together by silken threads, from which it drops, when the tree or branch is jarred, suspended in the air by a thread of silk. The larva is of a pale yellowish-green color, with a dusky or blackish stripe along each side, edged above by a narrow whitish stripe; there is also a dusky line along the middle of the back. Its head is shining yellow, and the top of the next segment is of the same color; on each ring there are several small black dots, from each of which arises a fine yellow hair. While young, the caterpillars eat only the green pulpy tissue of the leaves, leaving the network of veins entire; later on, they consume the whole of the leaf except its coarser veins. They also frequently gnaw holes or irregular cavities in the young apples. These larvæ feed on the leaves of the cherry as well as those of the apple.

When full grown, they are about half an inch long. They then change to chrysalids within the mass of eaten leaves occupied by the larvæ, and ordinarily spin a slight cocoon in a fold of a leaf, but when they are very abundant the foliage is so entirely consumed that they have to look for shelter elsewhere. Their chrysalids are then often found under dry leaves on the surface of the ground, in crevices in the bark of the tree, and in other suitable hiding-places. The chrysalis is about a quarter of an inch long; at first it is of a tawny-yellow color, which gradually changes to a darker hue. In ten or twelve days the perfect insect is produced.

Fig. 99.

The moth (Fig. 99) is of an ash-gray color. The fore wings are sprinkled with black atoms, and have four black dots near the middle, and six or seven smaller ones along the hinder margin. The hind wings are dusky above and beneath, with a glossy azure-blue reflection, blackish veins, and long, dusky fringes. The antennæ are alternately striped with black and white. Sometimes the fore wings are of a tawny yellow, in other specimens they are tinged with purplish red, and in some the dots are faint or entirely wanting. They rest with their long, narrow wings folded together and laid flat upon their backs.

Remedies.—Showering the trees with whale-oil soap and water has been recommended, but the use of Paris-green and water, as directed for No. 35, would prove more effectual; the water would dislodge many of the larvæ, and the remainder would be destroyed by eating the poisoned leaves.

In the year 1791 the orchards and forests of New England were overrun with this larva, and many of the trees perished. It was at that time that the insect received the popular name of Palmer-worm, which it has ever since retained. Another remarkable visitation occurred in 1853, which extended all over the Eastern States, and also over the eastern part of the State of New York. It was first observed about the middle of June, and so rapid was the destruction it occasioned that in a few days it was everywhere the leading topic of conversation and was generally regarded as a new and unknown insect. The trees attacked assumed a brown and withered appearance, looking as though they had been scorched by fire. Apple-trees and oaks suffered most, but nearly all other trees and shrubs were more or less injured. The weather was dry and hot previous to and during this period, but on the 20th of June copious rains fell, when the worms suddenly disappeared, the rain doubtless dislodging them, and perhaps drowning a large number of them. The fruit-crop in those sections that year was almost destroyed, from the trees losing their leaves by this insect. The following year they were quite scarce, and since then they have not appeared in such alarming numbers.

There are two other insects found on the apple-leaves resembling the Palmer-worm, and having similar habits, which are described by Dr. Asa Fitch as distinct, but which are probably varieties only of the common Palmer-worm. One of these is described as "the comrade Palmer-worm, *Chæto-chilus contubernalellus.*" The larva of this is found in company with the common Palmer-worm, from which it differs only in having the head and the upper part of the second segment of a polished black color. The moth of this black-headed larva differs from the common Palmer-worm moth chiefly in the ground-color of the wings, which are dark brown on the inner half, with the outer half white, the latter sometimes tinged with tawny yellowish. The other insect is described as "the tawny-striped Palmer-worm, *Chœtochilus malifoliellus*," and is a slender, pale-yellowish larva, similar in size to the ordinary Palmer-worm, with a tawny-yellow stripe along each side of the back, broadly margined above and below with white. The head is pale yellow, and there are a few minute dots scattered over the surface of the body, from each of which arises a fine hair. It appears during the early part of July, which is a little later than the common Palmer-worm, but has precisely similar habits. The moth is ash-gray and glossy, often with a purplish-red reflection, and differs from the moth of the common species in that the fore wings are not sprinkled with black atoms, and in having in addition to the dots on the fore wings a tawny-yellow band towards the tips, edged with whitish in front. Should these prove to be distinct and at any time troublesome, the treatment suggested for the common Palmer-worm will be equally applicable in either case.

No. 45.—Climbing Cut-worms.

These are the caterpillars of various night-flying moths, and are well known to horticulturists and gardeners everywhere. Most of the species are particularly destructive to young cabbage-plants and similar young and tender vegetation, cutting or severing the plants, when but three or four inches high, just above or below the ground, from which habit they derive their common name. They are active only at night, remaining concealed during the day just under the surface of the earth in the immediate neighborhood of their feeding-grounds. Some of the species are known as climbing cut-worms, and have the habit of ascending fruit-trees at night and committing great havoc among the expanding buds and young foliage, and it is to these that we here particularly refer.   Orchards having a light, sandy soil are much more liable to attack than those with a stiff and heavy soil.   Where the buds and foliage of trees or vines are being destroyed without apparent cause, climbing cut-worms should be searched for, when the lurking foes will usually be found buried in the soil not far from the base of the trees or vines injured.

The several species of climbing cut-worms, while differing in size, color, and markings, are much alike, being all smooth, naked larvæ of some shade of gray, green, brown, or black, with grayish or dusky markings.

FIG. 100.

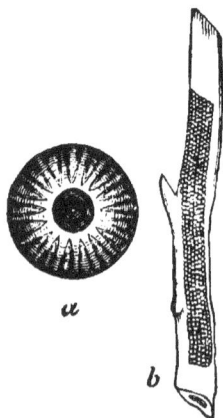

a

b

The Variegated Cut-worm, *Agrotis saucia* (Hubner).   One of the eggs of this species is represented in Fig. 100, much enlarged ; also a patch of the same, numbering several hundreds, on a twig.   The egg is round and flattened, of a pinkish color, and very prettily ribbed and ornamented.   These are often laid on twigs of the apple, cherry, and peach.

The young larvæ, when hatched, are very small, and of a dull-yellowish color, with darker spots.   At first, it is said, they do not hide themselves under the ground, but acquire this habit after their first moult, which takes place about a week after they are hatched.   They become full grown before the middle of June, when they present the appearance shown in Fig. 101, which shows the larva as at rest ; when extended and in motion, it is nearly two inches long.   The figure at the side represents the head magnified, showing its markings more distinctly. The full-grown caterpillar is of a dull flesh-color, mottled with brown and black, with elongated velvety black markings on each side.

When mature, the larva enters the ground, where it forms an oval, smooth cavity (see Fig. 102), within which it changes

FIG. 101.

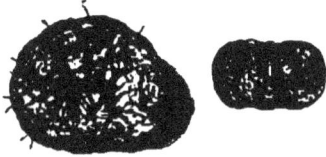

FIG. 102.

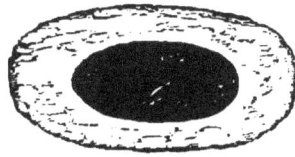

to a chrysalis of a deep mahogany-brown color, pointed at the extremity.

Within a few days the moth (Fig. 103) appears, which measures, when its wings are expanded, about an inch and three-quarters across. The fore wings are of a grayish-brown color, marked with brownish black; the hind wings are white and pearly, shaded towards the margin with pale brown.

FIG. 103.

The Dark-sided Cut-worm, *Agrotis Cochranii* Riley, is another of the climbing species. The caterpillar (*a*, Fig. 104) is a little over an inch in length, of a dingy ash-gray color above, much darker along the sides of the body. The chrysalis, which is formed under ground, is about seven-tenths of an inch long, of a yellowish-brown color, with darker brown markings. The moth is light gray, marked and shaded with brown.

FIG. 104.

a

The Climbing Cut-worm, *Agrotis scandens* Riley. The larva of this insect is a very active climber, and does a great deal of injury to fruit-trees. It is represented in Fig. 105 in the act of

Fig. 105.

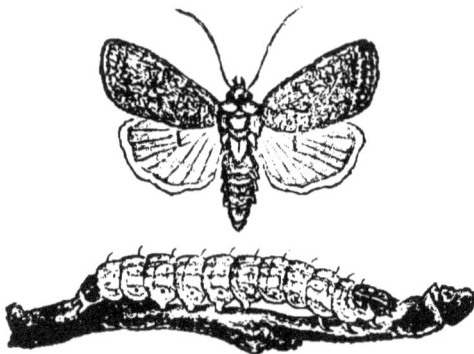

devouring the buds on a twig. It is of a light yellowish-gray color, variegated with dull green, with a dark line down the back, and fainter lines along the sides; the spiracles, or breathing-pores, are black. When full grown, it is nearly an inch and a half long, when it enters the earth, and there changes to a brown chrysalis. The moth (Fig. 105) has the fore wings of a light bluish gray, with darker markings, and the hind wings pearly white. The length of the body is about seven-tenths of an inch, and the wings measure, when spread, nearly an inch and a half across.

The W-marked Cut-worm, *Agrotis clandestina* (Harris) (Fig. 106), has also been found feeding on apple-buds, al-

Fig. 106.

though it more frequently attacks low bushes, such as currants; also succulent plants, such as young corn, cabbages, etc. The moth of this species (Fig. 107) has the fore wings of a rather dark ash-gray color, with the deeper lines and wavy bands but faintly traced. The hind wings are dull white, with a tinge of brown, becoming darker towards the hinder edge. The chrysalis is of the usual brown color, and is formed in a cell under the earth, as in the other species referred to.

The family of cut-worms is a large one, and embraces many other destructive species, but none of them, except those above mentioned, are known to have the habit of climbing trees. Some of the other injurious species will be referred to when treating of the insects which injure the strawberry.

Remedies.—One of the most effectual remedies against the climbing cut-worms is to fasten strips of tin or zinc around the tree, cut in such a way as to form, when applied, a sort of inverted funnel; this forms an effectual barrier to their ascent. They may also be collected by visiting the trees after dark and jarring or shaking them over sheets spread on the ground. It has also been suggested to dig holes about the trees, or on one side of them, with nearly perpendicular sides, when the cut-worms, being clumsy in their movements, are very likely to fall into them, and will not be able to get out again. Sprinkling the foliage with Paris-green or hellebore mixed with water, as recommended for No. 35, would no doubt poison them.

FIG. 107.

There are several parasites, both Ichneumons and Tachina flies, which attack cut-worms and greatly lessen their numbers. Some of the carnivorous beetles (see Figs. 47 and 48) also feed upon them.

No. 46.—The Lime-tree Winter-moth.

Hybernia tiliaria Harris.

The caterpillar of this species is a span-worm, not unlike the canker-worm, but larger and differently marked. The head is dull red, with a V-shaped mark on the front; the body yellow above, with many longitudinal black lines; the under side is paler. When full grown, it is about an inch and a quarter long. Besides the apple, it feeds on basswood, elm, and hickory. The larvæ hatch early in the spring, and sometimes prove very destructive to the foliage. In Fig. 108 they are represented both feeding and at rest. They complete their growth about the middle of June, when, letting themselves

down from the trees by a silken thread, they burrow into the
ground, forming a little oblong cell, five or six inches below
the surface, within which the change to a chrysalis takes
place, and from which the moth usually comes out late in

FIG. 108.

October or early in November, but occasionally this latter
change does not take place until spring.

The male moths have large and delicate wings (see Fig.
108) and feathered antennæ. The fore wings, which measure,
when expanded, about an inch and a half across, are of a
rusty-buff color, sprinkled with brownish dots, and with two
transverse wavy brown lines, the inner one often indistinct,
while between the bands and near the edge of the wing there
is generally a brown dot. The hind wings are paler, with a
small brownish dot in the middle; the body is similar in color
to the fore wings.

The female, also shown in Fig. 108, is a wingless, spider-like creature, with slender, thread-like antennæ, yellowish-white body, sprinkled on the sides with black dots, and with two black spots on the top of each ring except the last, which has only one. The head is black in front, and the legs are ringed with black. She is furnished with a jointed ovipositor, which can be protruded or drawn in at pleasure, and from which the eggs are deposited. As soon as the females leave the ground, they climb up the trees and await the attendance of the males.

The eggs are oval, of a pale-yellow color, and covered with a net-work of raised lines. They are laid in little clusters here and there on the branches.

As the habits of this insect are similar to those of the canker-worm, the remedies recommended for the latter will prove equally efficient in this instance.

No. 47.—The White Eugonia.

Eugonia subsignaria (Hubner).

This insect has only recently been reported as injurious to the foliage of the apple. It has long been known as destructive to shade-trees, particularly the elm. From a communication to the "Canadian Entomologist," vol. xiv. p. 30, by Mr. Charles R. Dodge, of Washington, D.C., it appears that the larva of this moth has become exceedingly injurious to apple-trees in some parts of Georgia.

Fig. 109.

The moth is pure white, and measures, when its wings are spread, about an inch and a half across. In the male the antennæ are pectinated or toothed (Fig. 109 represents a male); in the female they are much less toothed. When resting on the trees, these moths are easily disturbed, and on the slightest alarm drop to the ground for protection.

The eggs are usually deposited on the under side of the limbs, near the tops of the trees, in patches, consisting often of many hundreds, arranged in rows closely crowded together. They are smooth, irregularly ovoid, slightly flattened on the sides, rounded at the bottom, while the top is depressed, with a whitish rim or edge, forming a perfect oval ring. The egg hatches about the 1st of May.

The caterpillar (Fig. 110) is dark brown, with a large red head; the terminal segment is also red. It lives in this stage about forty days, and then changes

FIG. 110.

to a chrysalis, in which condition it remains about ten days, when the moth escapes. This insect, when very abundant, devours the leaves of almost every variety of tree, bush, and shrub.

Where abundant, they may be poisoned, and the orchard protected, by syringing the trees with Paris-green and water, in the proportion of a teaspoonful of the poison to two gallons of water.

No. 48.—The Hag-Moth Caterpillar.

Phobetron pithecium (Sm. & Abb.).

The caterpillar of this moth is a curious, slug-like creature, of a dark-brown color, flattened, oblong, or nearly square in

FIG. 111.

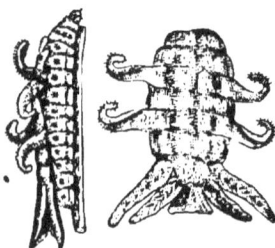

form, with singular, fleshy appendages protruding from the sides of its body. The three middle ones are longest, measuring about half an inch long, and have their ends curved. When this larva is handled, the fleshy horns become detached, and when spinning its cocoon it detaches them and fastens them to the outside. Fig. 111 gives a side view as well as a back view of this larva. It feeds on the cherry as well as the apple.

The cocoon is small, round, and compact, usually fastened to a limb or twig of the tree on which the larva has fed.

The moth escapes in about ten days. It is of a dusky-brown color, the front wings variegated with pale yellowish brown, and crossed by a narrow, wavy, curved band of the same color, edged near the outer margin with dark brown, and having near the middle a light-brown spot. When its wings are expanded, it measures from an inch to an inch and a quarter across. It is an insect which has always hitherto been rare, and is never likely to do much injury.

No. 49.—The Saddle-back Caterpillar.

Empretia stimulea Clemens.

This caterpillar, which is represented in Fig. 112, *a*, a back view, *b*, a side view, is often found feeding on apple-leaves, also on those of the cherry, grape, raspberry, currant, rose, althæa, Indian corn, and sumach. It is of a reddish-brown color, rounded above, flattened beneath, armed with prickly thorns, which are longest on the fourth and tenth segments, and with a bright pea-green patch, somewhat resembling a saddle in

FIG. 112.

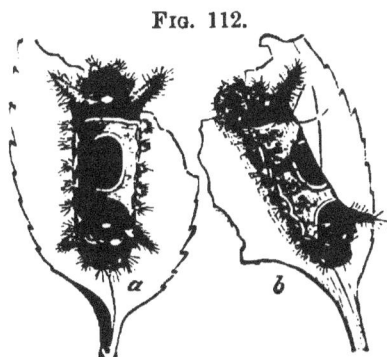

form, over the middle portion of the body, centred with a broad, elliptical, reddish spot, the red spot and green patch both being edged with white. The thorns with which the body is armed sting like a nettle when applied to the back of the hand, or any other part where the skin is tender, and the parts touched swell with watery pustules, the irritation being accompanied with much itching. The under part of the body of the larva is flesh-colored; there are three pairs of thoracic legs, but the thick, fleshy, abdominal legs found in most other caterpillars are wanting in this species, and the larva glides along with a snail-like motion.

The cocoon is rounded, almost spherical, and is surrounded with a loose silken web.

The moth (Fig. 113) appears on the wing from the middle to the end of June; but it is a rare insect, and is seldom captured even by collectors. The wings are of a deep, rich, reddish, velvety brown, with a dark streak about the middle of the fore wings, extending from the body half-way across, and on this is a golden spot; there are also two golden spots near the apex of the wing. When the wings are spread they measure nearly an inch and a half across.

Fig. 113.

In the larval state this insect is preyed on by a small Ichneumon fly, and, never being abundant, other remedies are not needed to subdue it.

No. 50.—The Apple-leaf Miner.

Tischeria malifoliella Clemens.

The larva of this insect lives within the leaf of the apple-tree, between the upper and the under skin, devouring the soft tissues, and burrowing an irregular channel, which begins as a slender white line, dilating as the larva increases in size, and ultimately becoming an irregular brownish patch, sometimes extending to, or over, the place of beginning. The caterpillar is of a pale-green color, with a brown head, and the next segment brownish.

When about to change to a pupa, the leaf is drawn into a fold, which is carpeted with silk, and in this enclosure the chrysalis is formed, the change occurring during September. When the leaf falls, its occupant falls with it, and remains on the ground within the folded leaf until the following May.

The moth is a tiny creature, measuring, when its wings are spread, a little more than a quarter of an inch across. The

fore wings are of a shining dark brown, suffused with a tinge
of purple, and slightly dusted with dull-yellowish atoms. The
hind wings are dark gray.

This insect also mines the leaves of the wild crab-apple,
different species of thorn, the blackberry, and the raspberry,
but has never been known to do any material injury.

No. 51.—The Apple-tree Case-bearer.

Coleophora malivorella Riley.

With the opening of spring there will sometimes be found
on the twigs of apple-trees curious little pistol-shaped cases as
shown at *a*, Fig. 114. Each of these on examination will be

FIG. 114.

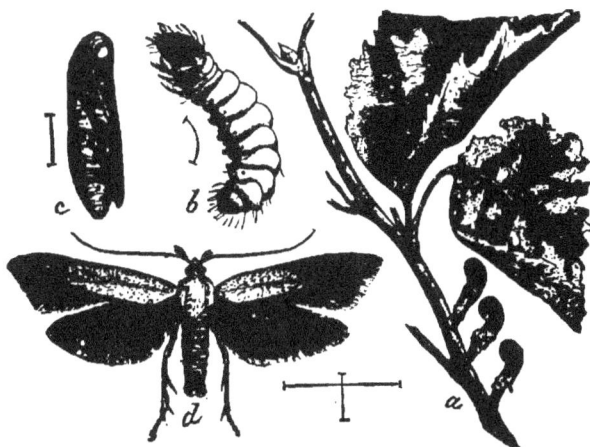

found to contain a larva, possessing the power of moving from
place to place and carrying its protecting case with it. These
cases are very tough, almost horny in their texture, and seem
to be proof against the attack of insect enemies. As the buds
begin to swell, the cases will be found here and there sticking
on them, while the active little foe within is busily devouring
their interior. In this way many of the fruit-buds are destroyed, nothing but hollow shells being left. As the season
advances, the caterpillars leave the twigs and fasten on the

leaves, on which they also feed, sometimes reducing them to mere skeletons. Late in June the change to chrysalis takes place, and the moths appear on the wing in July. They fly at night, and deposit their eggs on the leaves; these eggs hatch during August and September, the larvæ living and feeding on the under side of the leaves until frost comes, when before the leaves fall they migrate to the twigs, and, fastening their odd little cases firmly with silken threads, remain torpid until the following spring; then, aroused to activity by the first warm days, they attack the swelling buds, as already described.

The larva (b, Fig. 114) is of a pale-yellow color, with a faint rosy tint, a black head, and a few short hairs on its body. In the figure it is much magnified; the hair-line adjoining shows its natural size; c represents the chrysalis, and d the moth, both enlarged. The wings of the moth are brown, with white scales, head and thorax white, abdomen whitish, all dotted with brown scales. The wings, when expanded, measure a little more than half an inch across.

No. 52.—The Resplendent Shield-bearer.

Aspidisca splendoriferella Clemens.

Occasionally there may be found on the limbs of apple-trees during the winter clusters of little oval seed-like bodies, as shown at d, Fig. 115; these on examination will be found to be formed of minute portions of apple-leaves, and on opening one of them it will be seen to contain a small yellowish larva, or, if the season be advanced, perhaps a chrysalis.

During the month of May a very small but very beautiful moth escapes from each of these enclosures. The moth is represented at g in Fig. 115, much magnified. Its head is golden, the antennæ brown, tinged with gold; the fore wings from the base to the middle are of a leaden gray with a metallic lustre, and from the middle to the tip golden; a broad silvery streak extends from the front edge to about the middle, margined with a dark color on both sides; there are also other streaks and spots of silvery and dark brown. The hind wings are

Syn. Coptodisca splendoriferella
Kentucky Bulletin no. 133.

f a rich deep gray margined with a long yellowish-brown
·inge. It is an active little creature, running about on the
pper surface of the leaves in the sunshine, with its wings
losely folded to its body.

The eggs are laid on the apple-leaves, and the young larva

Fig. 115.

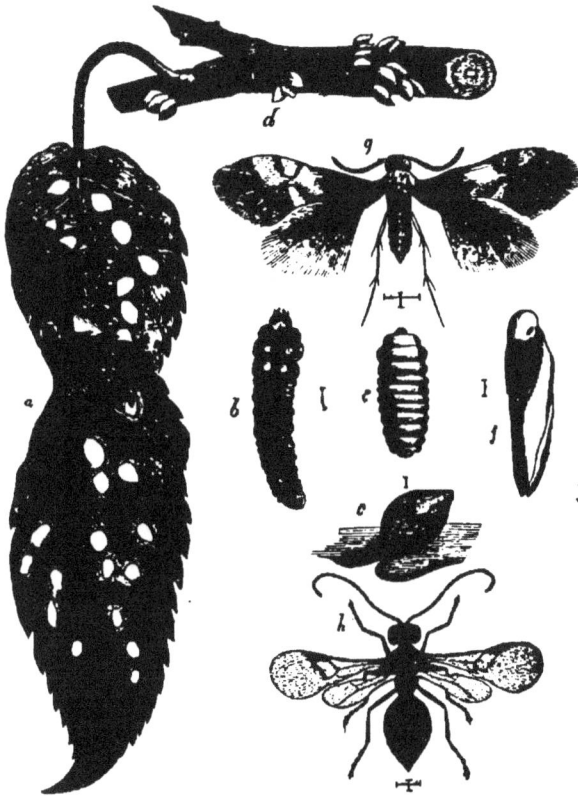

·hen hatched penetrates to the interior of the leaf, mining it,
·eaving the upper and under surfaces unbroken, but forming
·fter a time an irregular, dark-colored blotch upon the leaf.
·When mature, it forms from the leafy blotch its little case,
·nd, crawling with it, fastens it securely to a near twig or
·ranch of the tree. At this period the larva presents the

appearance shown at *b*, and is then about one-eighth of an inch long, and of a yellowish-brown color, with a dark head. Shortly, contracting within its case, it appears as shown at *c*, and finally transforms to a chrysalis, as seen at *f* in the figure.

There are two broods during the season, the moths appearing in May and again in July and August, the first brood of the larvæ being found in June, the second brood at the latter end of the season.

Remedies.—A minute parasitic fly, shown at *h* in Fig. 115, attacks this tiny creature and destroys it. (All these figures, except that of the leaf, are much magnified, the short lines at the side or below showing the natural size.) Should these insects prevail to such an extent as to require man's interference, the cases might be scraped from the branches and destroyed during the winter, or the limbs brushed with the alkaline wash or the mixture of sulphur and lime recommended for the woolly apple-louse, No. 9.

No. 53.—The Apple-leaf Bucculatrix.

Bucculatrix pomifoliella Clemens.

The larva of this insect feeds externally on the leaves of apple-trees, and is very active, letting itself down from the tree by a silken thread when disturbed. When full grown, it is nearly half an inch long, with a brown head and a dark yellowish-green body, its anterior portion tinged with reddish, and having a few short hairs scattered over its surface.

When full grown, the caterpillar spins an elongated, whitish cocoon, attached to the twig on the leaves of which it has been feeding; this cocoon is ribbed longitudinally, as shown at *b*, Fig. 116, and within this enclosure the larva changes to a brown chrysalis. The second brood is found late in the autumn, the insect remaining in the chrysalis state during the winter. The moths issue the following spring, when they lay eggs for the first brood of caterpillars, which are found injuring the foliage during the month of June.

The fore wings of the moth (*c*, Fig. 116) are whitish,

tinged with pale yellow, and dusted with brown. On the middle of the inner margin is a large, oval patch of dark brown, forming, when the wings are closed, a conspicuous, nearly round spot; there is a wide streak of the same hue opposite, extending to the front margin, and a dark-brown spot near the tip. In the figure the moth is shown

FIG. 116.

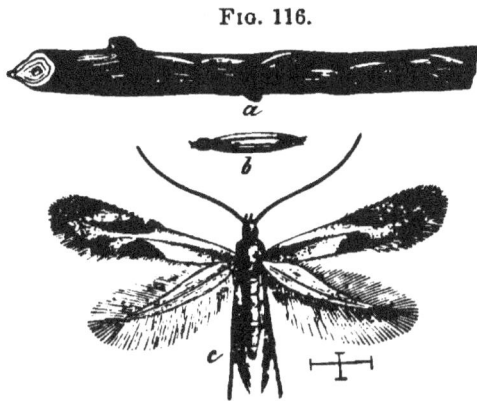

highly magnified. Sometimes this insect appears in immense numbers, and then becomes injurious.

Remedies.—As the cocoons of the second brood remain attached to the trees all winter, abundant opportunity is afforded to destroy them. Any oily or alkaline liquid brushed over them will usually penetrate and destroy the enclosed insect. A minute parasitic fly is destructive to this pest, and the cocoons may often be found perforated with small round holes at one end, through which these tiny friends have escaped.

No. 54.—The Apple Lyonetia.

Lyonetia saccatella Packard.

This is a tiny moth, but a very beautiful one, which appears early in the summer; its wings, when expanded, measure only one-fifth of an inch across. It is shown, much magnified, in Fig. 117. The fore wings are of a light slate-gray on the inner half, while the outer half is bright orange, enclosing two white bands, one arising on the front edge, the other on the inner margin, both nearly meeting in the middle of the wing; these white bands are margined externally with black.

FIG. 117.

There is a conspicuous black spot near the fringe, from which arises a pencil of black hairs.

The larva (Fig. 118), which feeds on apple-leaves, is small, flattened, and of a green color. It constructs from the skin of the leaf a flattened, oval case, in which it lives; the case is open at each end, and is drawn about by the larva as it moves from place to place. The case is represented in Fig. 119. (Both case and larva are magnified.) The larva becomes full grown about the end of August, and attaches its cocoon to the bark of the tree on which it is feeding, changing there to a chrysalis, in which condition it remains until the following spring.

FIG. 118.   FIG. 119.

No. 55.—The Rosy Hispa.

Odontota rosea (Weber).

This is a small, flat, rough, coarsely-punctated beetle, its wing-covers forming an oblong square, as shown in Fig. 120; there are three smooth, raised, longitudinal lines on each of them, spotted with red, while the spaces between are deeply punctated with double rows of dots. The head is small, the antennæ short, thickened towards the end, and the thorax rough above, striped with deep red on each side. The under side of the body is usually darker in color, sometimes blackish. This beetle is found from the latter part of May until the middle of June, and deposits its eggs on the leaves of the apple-tree. These are small, rough, and of a blackish color, fastened to the surface of the leaves, sometimes singly and sometimes in clusters of four or five.

FIG. 120.

The larvæ, when hatched, eat their way into the interior of the leaf, where they feed upon its green, pulpy substance, leaving the skin above and below entire, which soon turns brown and dry, forming a blister-like spot. The larva, when full grown, which is usually during the month of July, is about one-fifth of an inch long, oblong in form, rather broader before than behind, flattened, soft, and of a yellowish-white color, with the head and neck blackish and of a horny consistence. Each of the three anterior segments has a pair of legs; the other segments are provided with small fleshy warts at the sides, and transverse rows of little rasp-like points above and beneath.

The larva changes to a pupa within the leaf, from which, in about a week, the perfect insect escapes. Within these blister-like spots the larva, pupa, or freshly-transformed beetle may often be found. This insect never occurs in sufficient numbers to be a source of much trouble.

No. 56.—The Cloaked Chrysomela.

Glyptoscelis crypticus (Say).

This is another beetle which devours the foliage of the apple-tree, also that of the oak-tree. It is of a thick, cylindrical form, about one-third of an inch long, with its head sunk into the thorax, and the thorax narrower than the body. It is of a pale ash-gray color, from being Fig. 121. entirely covered with short whitish hairs. The closed wing-covers have a small notch at the top of their suture. At the junction of the wing-covers with the thorax there is a dusky spot. This insect is represented in Fig. 121.

No. 57.—The Apple-tree Aphis.

Aphis mali Fabr.

During the winter there may often be found in the crevices and cracks of the bark of the twigs of the apple-tree, and also about the base of the buds, a number of very minute, oval, shining black eggs. These are the eggs of the apple-tree aphis, known also as the apple-leaf aphis, *Aphis mali-foliæ* Fitch. They are deposited in the autumn, and when first laid are of a light yellow or green color, but gradually become darker, and finally black.

As soon as the buds begin to expand in the spring, these eggs hatch into tiny lice, which locate themselves upon the swelling buds and the small, tender leaves, and, inserting their beaks, feed on the juices. All the lice thus hatched at this period of the year are females, and reach maturity in ten or twelve days, when they commence to give birth to living young, producing about two daily for two or three weeks, after which the older ones die. The young locate about the parents as closely as they can stow themselves, and they also mature and become mothers in ten or twelve days, and are as prolific as their predecessors. They thus increase so rapidly that as fast as new leaves expand colonies are ready to occupy them. As the season advances, some of the lice acquire wings, and, dispersing, found new colonies on other trees. When cold weather approaches, males as well as females are produced, and the season closes with the deposit of a stock of eggs for the continuance of the species another year.

When newly born, the apple aphis is almost white, but soon becomes of a pale, dull greenish-yellow. The females are said to be always wingless; their bodies are oval in form, less than one-tenth of an inch long, of a pale yellowish-green color, often striped with deeper green. The eyes are black, honey-tubes green, and there is a short, tail-like appendage of a black color. The accompanying illustration (Fig. 122) of a winged male and wingless female, highly magnified, shows the structure and shape of the insect; its beak, which proceeds from

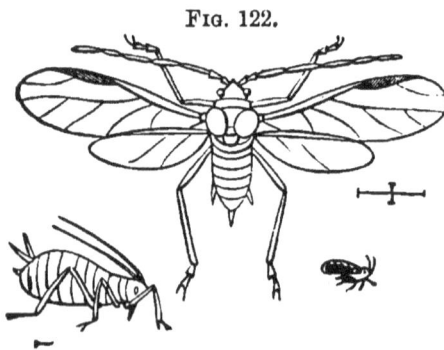

Fig. 122.

the under side of the head, is here hidden from view in the male, but can be seen in the female.

Both the winged and wingless lice are very similar in color. The head, thorax, and antennæ are black, with the neck usually green. The abdomen is short and thick, of an oval form and bright-green color, with a row of black dots along each side; the nectaries and tail-like appendage are black; the wings are transparent, with dark-brown veins.

Most of the insects belonging to this family are provided with two little tubes or knobs, which project, one on each side, from the hinder part of their bodies; these are called honey-tubes, or nectaries, and from them is secreted in considerable quantities a sweet fluid. This fluid falling upon the leaves and evaporating gives them a shiny appearance, as if coated with varnish, and for the purpose of feeding upon this sweet deposit, which is known as honey-dew, different species of ants and flies are found visiting them. Ants also visit the colonies of aphides and stroke the insects with their antennæ to induce them to part with some of the sweet liquid, which is greedily sipped up. This fluid is said to serve as food for a day or two to the newly-born young.

The leaves of trees infested by these insects become distorted and twisted backwards, often with their tips pressing against the twig from which they grow, and they thus form a covering for the aphides, protecting them from rain. An infested tree may be distinguished at some distance by this bending back of the leaves and young twigs. It is stated that the scab on the fruit of the apple-tree often owes its origin to the punctures of these plant-lice. This species, which was originally imported from Europe, is now found in apple-orchards all over the Northern United States and Canada.

Remedies.—Scraping the dead bark off the trees during the winter and washing them with a solution of soft soap and soda, as recommended for No. 2, the two-striped borer, would be beneficial, by destroying the eggs. Syringing the trees, about the time the buds are bursting, with strong soap-suds, weak lye, or tobacco-water, the latter made by boiling one pound of the rough stems or leaves in a gallon of water, will destroy a large number of the young lice. A frost occurring after a few days of warm weather will kill millions of them; in the egg state the insects can endure any amount of frost, but the young aphis quickly perishes when the temperature falls below the freezing-point.

Myriads of these aphides are devoured by Lady-birds and their larvæ. In Fig. 123 is represented the Nine-spotted

FIG. 123.	FIG. 124.	FIG. 125.

Lady-bird, *Coccinella novemnotata* Herbst, one of our commonest species, which is found almost everywhere; it is of a brick-red color, and is ornamented with nine black spots.

The Two-spotted Lady-bird, *Adalia bipunctata* (Linn.) (Fig. 124), is also extremely common. This is very similar in color to the nine-spotted species, but in this one there is only a single spot on each wing-case. In the figure the insect is shown magnified.

Fig. 125 represents the Plain Lady-bird, *Cycloneda sanguinea* (Linn.). This is somewhat smaller in size than the last two species named, of a lighter shade of red, and without any spots on its wing-cases. It is known also as *Coccinella munda.*

The Comely Lady-bird, *Coccinella venusta* Mels. (Fig. 126), is pink, with ten large black spots, the hinder ones being united together.

The Thirteen-spotted Lady-bird, *Hippodamia 13-punctata* (Herbst), is shown in Fig. 127; it is larger than *C. sanguinea*, and has thirteen black spots on a brick-red ground.

In Fig. 128, *c*, is represented the Convergent Lady-bird,

Hippodamia convergens Guer., which is of an orange red, marked with black and white. The larva is shown of its

Fig. 126.   Fig. 127.   Fig. 128.

  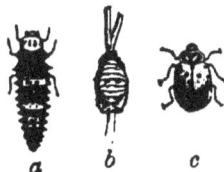

a   b   c

natural size at *a*, its colors being black, orange, and blue, and when full grown it attaches itself to the under side of a leaf and changes to a pupa, which is shown at *b*.

The Spotted Lady bird, *Megilla maculata* (De Fig. 129. Geer) (see Fig. 129), is of a pinkish color, some-  times pale red. It has large black blotches, twelve in all, on its wing-cases; two on one wing-cover are opposite to and touch two on the other.

Fig. 130 represents the Fifteen-spotted Lady-bird, *Anatis 15-punctata* (Oliv.), the largest of them all. It is a very

Fig. 130.

variable insect; at *d, e, f, g*, are shown four of the different forms under which it is seen ; *a* shows the larva in the act of devouring a young larva of the Colorado potato-beetle, to which it is also partial, while *b* represents the pupa.

The Painted Lady-bird, *Harmonia picta* (Rand), is a very pretty little insect. (See Fig. 131.) At *b* it is shown of the natural size, at *c* enlarged ; it is of a pale straw-color, marked with black, as in the figure. The larva, *a*, is of a dusky brown, with paler markings.   This species is most commonly found feeding on lice which attack the pine.

All the Lady-birds are very useful creatures, and, with their

Fig. 131.

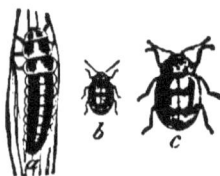

Fig. 132.

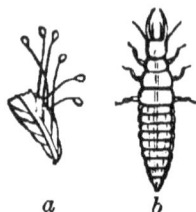

a   *b*

larvæ, should be encouraged and protected by the fruit-grower in every possible way.

The larvæ of the Lace-winged or Golden-eyed Flies, *Chrysopa*, are equally destructive to aphides, roaming about among them like so many tigers with appetites almost insatiable.   At *b*, Fig. 132, one of these larvæ is shown, and at *a* some of the eggs, which are attached to the end of fine upright threads or stalks.   These are usually found in clusters.   The perfect in-

Fig. 133.

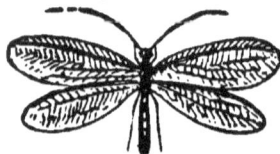

Fig. 134.

sect has four delicate, transparent, whitish wings (see Fig. 133) netted like fine lace, bright-golden eyes, and a beautiful green body.   Fig. 134 shows the same insect with its wings closed ; also a side view of a cluster of eggs.   While beau-

Fig. 135.

tiful to look at, these insects are offensive to handle, as when touched they emit a very sickening, pungent, and persistent odor.

Other friendly helpers in this good work are the larvæ of the Syrphus flies.   These are fleshy larvæ, thick and blunt behind, and pointed in front.   (See Fig. 135.) Their mouths are furnished with a triple-pointed dart, with which they seize and pierce their prey, and, elevating it, as shown in the figure, deliberately suck it dry. They are quite blind, but the eggs from which they hatch are deposited by the parent flies in the midst of the colonies of plant-lice, where they grope about and obtain an abundance of food without much trouble. In Fig. 136 is shown one of the flies. They are black with transparent wings, and are prettily ornamented with yellow stripes across their bodies.

Fig. 136.

ATTACKING THE FRUIT.

No. 58.—The Codling Moth.

Carpocapsa pomonella (Linn.).

In the accompanying figure, 137, *a* shows the burrowings of this larva, *b* the point where it effected its entrance, *e* the larva full grown, *h* the anterior part of its body, magnified, *d* the chrysalis, *i* the cocoon, *f* the moth with its wings closed, and *g* the same with wings expanded. A better representation of the moth is given, magnified, in Fig. 138. The larger opening at the side of the apple shows where the full-grown larva has escaped.

This is one of the most troublesome insects with which fruit-growers have to contend, and although of foreign origin, having been im-

Fig. 137.

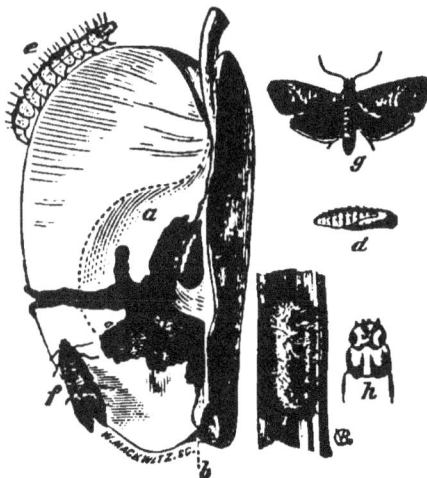

ported from Europe about the beginning of the present cen-
tury, it is now found in almost all parts of North America,
entailing an immense yearly loss upon apple-growers.

The early brood of moths appear on the wing about the
time of the opening of the apple-blossoms, when the female
deposits her tiny yellow eggs singly in the calyx or eye, just
as the young apple is forming ; in a few instances they have
been observed in the hollow at the stalk
end, and occasionally on the smooth
surface of the cheek of the apple.   In
about a week the egg hatches, and the
tiny worm at once begins to eat through
the apple to the core.   Usually its castings are pushed out through the hole
by which it has entered, the passage being enlarged from
time to time for this purpose. Some of the castings commonly
adhere to the apple ; hence, before the worm is full grown,
infested fruit may generally be detected by the mass of reddish-brown exuviæ protruding from the eye.   Sometimes as
the larva approaches maturity it eats a passage through the
apple at the side, as shown in the figure, and out of this
opening thrusts its castings, and through it the larva, when
full grown, escapes.   The head and upper portion of the first
segment of the young larva are usually black, but as it approaches maturity these change to a brown color.   The body
is of a flesh-color, or pinkish tint, more highly colored on
the back ; it is also sprinkled with minute, elevated points,
from each of which there arises a single fine hair.

In three or four weeks from the time of hatching the early
brood of larvæ attain full growth, when the occupied apples
generally fall prematurely to the ground, sometimes with the
worm in them, but more commonly after it has escaped.   The
larvæ, which leave the apples while still on the trees, either
crawl down the branches to the trunk of the tree, or let themselves down to the ground by a fine silken thread, which they
spin at will.   In either case, whether they crawl up or down,

Fig. 138.

the greater portion of them find their way to the trunks of the trees, where, under the rough bark and in cracks and crevices, they spin their cocoons.

Having selected a suitable hiding-place, the larva constructs a papery-looking silken cocoon, shown at *i* in the figure, which is white inside, and disguised on the outside by attaching to the silky threads small fragments of the bark of the tree or other available débris. After the cocoon is completed, the change to the chrysalis takes place in the early brood in about three days. At first the pupa is of a pale-yellow color, deepening in a day or two to pale brown; the insect remains in this condition about two weeks, when the moth escapes.

Each moth is capable of laying on an average probably not less than fifty eggs, but these are not all matured at once; by careful dissection they may be found in the body of the moth in different stages of development. Hence they are deposited successively, extending over a period probably of from one to two weeks or more; add to this the fact that some of the moths are retarded in their development in the spring, and it is easy to account for the finding of larvæ of various sizes at the same time; indeed, sometimes the later specimens from the first brood will not have escaped from the fruit before some of the young larvæ of the second brood make their appearance, the broods thus, as it were, overlapping each other, and very much extending the period for the appearance of the winged insects.

The moth (*g*, Fig. 137), although small, is a beautiful object. The fore wings are marked with alternate irregular, transverse, wavy streaks of ash-gray and brown, and have on the inner hind angle a large, tawny-brown spot, with streaks of light bronze or copper color, nearly in the form of a horseshoe; at a little distance they resemble watered silk. The hind wings and abdomen are of a light yellowish brown, with the lustre of satin. The moth conceals itself during the daytime, and appears only at night, and, since it is not readily attracted by light, is seldom seen. The second brood of

9

moths are usually on the wing during the latter half of July, when they pair, and in a few days the female begins to deposit her eggs for the later brood of larvæ, generally selecting for this purpose the later apples. These larvæ mature during the autumn or early winter months; if they escape before the fruit is gathered, they seek some sheltered nook under the loose bark of a tree or other convenient hiding-place; but if carried with the fruit into the cellar, they may often be found about the barrels and bins in which it is stored; a favorite hiding-place is between the hoops and staves of the apple-barrels, where they are found sometimes by hundreds. If thus provided with snug winter-quarters, and through negligence allowed to escape, the fruit-grower must expect to suffer increased loss from his want of care. Having fixed on a suitable spot, the larva spins its little tough cocoon, firmly fastened to the place of attachment, and within this it remains in the larval state until early the following spring, when it changes to a brown chrysalis, and shortly afterwards the moth appears, to begin the work of the opening season.

Besides injuring the apple, it is very destructive to the pear; it is also found on the wild crab, and occasionally on the plum and peach.   Sometimes two larvæ will be found in the same fruit.

Remedies.—One of the most effective methods yet devised for reducing the numbers of this insect is to trap the larvæ and chrysalids and destroy them.   This is best done by applying bands around the trunks of the trees about six inches in width; strips of old sacking, carpet, cloth, or fabric of any kind will serve the purpose, and, although not so durable, many use common brown paper.   Whatever material is used, it should be wound entirely round the tree once or twice, and fastened with a string or tack.   Within such enclosures the larvæ hide and transform.   The bands should be applied not later than the 1st of June, and visited every eight or ten days until the last of August, each time taken off and examined, and all the worms and chrysalids found under them destroyed; they should also be visited once after the crop is secured. Some persons prefer to use narrower bands, not more than four inches wide, and fasten them with a tack, while others secure them in their place by merely tucking the end under. Usually the cocoons under the bandages are partly attached to the tree and partly to the bandage, so that when the latter is removed the cocoon is torn asunder, when it often happens that the larva or chrysalis will fall to the ground, and, if it escapes notice, may there complete its transformations. Wide-mouthed bottles partly filled with sweetened water, and hung in the trees, have been recommended as traps for the codling moth, but there is no evidence that any appreciable benefit has ever been derived from their use. A large number of moths can be captured in this manner, but it is rare to find a codling moth among them. Neither is the plan of lighting fires in the orchard of much avail, since codling moths are rarely attracted by light. Spraying the trees soon after the fruit has set, and while it is still in an upright position, with a mixture of Paris green and water in the proportion of a teaspoonful to a pailful of water, will deter the moths from placing their eggs on the apples, and thus protect much of the fruit from injury.

The fallen fruit should be promptly gathered and destroyed. It has been recommended that hogs be kept in the orchard for the purpose of devouring such fruit; and, where they can be so kept without injury to the trees or to other crops, they will no doubt prove useful.

FIG. 189.

This insect, while in the larval state, is so protected within the apple that it enjoys great immunity from insect enemies. Nevertheless it is occasionally reached by the ever-watchful Ichneumons, two species

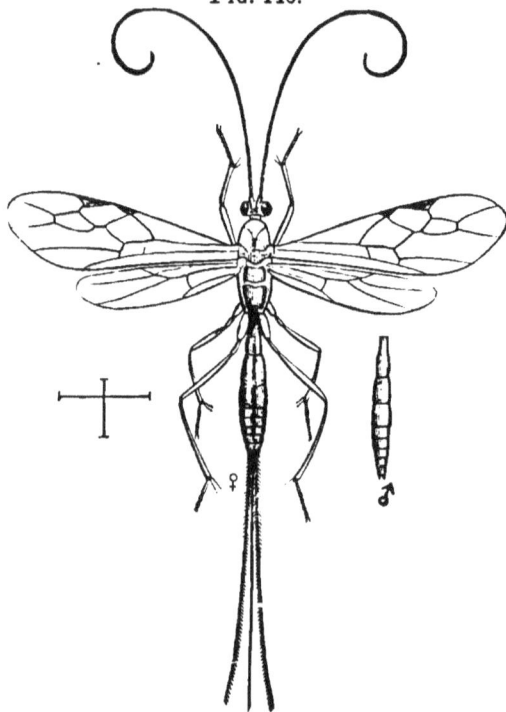

of which are known to occur as parasites within the bodies of
the larvæ.   They have been bred by Mr. C. V. Riley, who
describes them in his fifth Missouri Report.   One is a small
black fly, from one-fourth to one-half inch in length ; its legs
are reddish, the hind pair having a broad white ring.   It
is called the Ring-legged Pimpla, *Pimpla annulipes* Br., and
is represented, much magnified, in Fig. 139.   The other
species is about the
same size, but more
slender, and of a
yellow or brownish-
yellow color.   The
female is provided
with a long ovipos-
itor, as seen in
Fig. 140, where the
insect is shown
highly magnified.
The abdomen of the
male is represented
to the right of the
figure.   This spe-
cies is known as
the Delicate Long-
sting, *Macrocentrus
delicatus* Cresson.
These useful insect
friends are not yet
sufficiently numer-
ous to check materially the increase of the codling moth,
and it is doubtful if they ever will be.   When the codling
worm has left the fruit in which it has been feeding, and while
wandering about in search of a suitable spot in which to pass
its chrysalis stage, it is liable to be attacked by any of the
ground-beetles, *Carabidæ*, both in their larval and their
perfect state, also by the larvæ of soldier-beetles and other

Fig. 140.

carnivorous insects. Some of the smaller insectivorous birds are also said to devour this insect both in the larval and in the pupal condition.

No. 59.—The Apple Curculio.
Anthonomus quadrigibbus Say.

This is a small beetle, a little smaller than a plum curculio, of a dull-brown color, having a long, thin snout, which sticks out more or less horizontally, and cannot be folded under the body, as is the case with many species of Curculio. This snout in the female is as long as the body; in the male it is about half that length. In addition to the prominent snout, it is furnished with four conspicuous brownish-red humps towards the hinder part of its body, from which it takes its specific name, *quadrigibbus.* Including the snout, its length is a quarter of an inch or more. In the accompanying figure, 141, the insect is magnified; *a* represents a back view, *b* a side view; the outline at the left shows its natural size. Its body is dull brown, shaded with rusty red; the thorax and anterior third of the wing-covers are grayish.

FIG. 141.

b   *a*

This is a native American insect which formerly bred exclusively in the wild crabs and haws; it is single-brooded, and passes the winter in the beetle state. The beetle appears quite early, and the larva may often be found hatched before the middle of June, and in various stages of its growth in the fruit during June, July, and August.

The beetle with its long snout drills holes into the young apples, much like the puncture of a hot needle, the hole being round, and surrounded by a blackish margin. Those which are drilled by the insect when feeding are about one-tenth of an inch deep, and scooped out broadly at the bottom;

those which the female makes for her eggs are scooped out still more broadly, and the egg is placed at the bottom. The egg is of a yellowish color, and in shape a long oval, being about one-twenty-fifth of an inch in length and not quite half that in width. As soon as the larva hatches, it burrows to the heart of the fruit, where it feeds around the core, which becomes partly filled with rust-red excrement. In about a month it attains full size, when it presents the appearance shown in Fig. 142; *b* represents the larva highly magnified, and *a* the pupa.

The larva is a soft, white grub, nearly half an inch in length, with a yellowish-brown head and jaws. Its body is

Fig. 142.

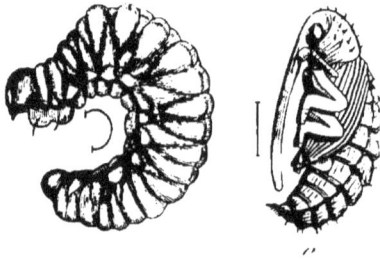

much wrinkled, the spaces between the folds being of a bluish-black color; there is also a line of a bluish shade down the back. Having no legs, it is incapable of much movement, and remains within the fruit it occupies, changing there to a pupa of a whitish color (see Fig. 142 *a*), and in two or three weeks, when perfected, the beetle cuts a hole through the fruit and escapes.

When feeding, this insect makes a number of holes or punctures, and around these a hard knot or swelling forms, which much disfigures the fruit; pears, as well as apples, are injured in this way. The infested fruits do not usually fall to the ground, as do apples affected by the codling worm, but remain attached to the tree, and the insect, from its habit of living within the fruit through all its stages, is a difficult one to destroy. Picking the affected specimens from the tree, and vigorously jarring the tree during the time when the beetle is about, will bring it to the ground, where it can be destroyed in the same manner as recommended for the plum curculio. Fortunately, it is seldom found in such abundance as to do much damage to the fruit-crop. In Southern Illinois and in some portions of Missouri it has proved destructive, but in most of the Northern United States and in Canada, although common on thorn-bushes and crab-apples, it seldom attacks the more valuable fruits to any considerable extent.

No. 60.—The Apple Maggot.

Trypeta pomonella Walsh.

This is a footless maggot, shown at *a*, Fig. 143, tapering to a point in front, and cut squarely off behind, which lives in the pulp of the apple, and tunnels it with winding channels, making here and there little roundish discolored excavations about the size of a pea. This maggot is of a greenish-white color, about one-fifth of an inch long, with a pointed head and a pale-brown, flattish, rough tubercle behind it; the hinder segment has two pale-brown tubercles below.

The pupa is of a pale yellowish-brown color, and differs from the larva only in being contracted in length; in this instance the true pupa is enclosed within the shrunken skin of the larva. When about to change, the maggot leaves the apple, and, falling to the ground, burrows under the surface, and there enters the pupal state, in which condition it remains until the middle of the following summer, when the perfect insect escapes in the form of a two-winged fly.

The fly (*b*, Fig. 143) is about one-fifth of an inch long, and measures, when its wings are expanded, nearly half an inch across. The head and legs are rust-red, the thorax shining black, more or less marked with grayish or white; the abdomen is black, with dusky hairs, and with whitish hairs bordering the spaces between the segments of the body. The wings are whitish glassy, with dusky bands. This insect is single-brooded, the fly appearing in July, when, by means of a sharp ovipositor, it inserts its eggs into the substance of the apple. It frequently attacks apples which have been previously perforated by the codling worm, and it prefers the thin-skinned summer and fall apples to the winter varieties It is, however, frequently found in apples which have been stored, and has thus proved very troublesome in many parts

FIG. 143.

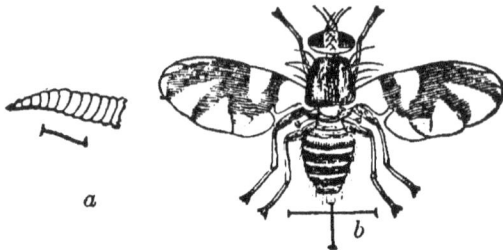

of the country, especially in Massachusetts, Connecticut, and New York. It is a native insect, found feeding on haws, and probably also on crab-apples.

No. 61.—The Apple Midge.

Sciara mali (Fitch).

This is also a small maggot, found devouring the flesh of ripened and stored apples, and hastening their decay. It appears to attack chiefly, if not wholly, those specimens which have been previously perforated by the codling worm, thus adding to the damage caused by that destructive pest, and when this insect has completed its transformations within the apple, the hole made by the codling worm affords this fly a ready means of exit.

The larvæ are long and slender, tapering gradually to a point at the head, the hinder end being blunt; they are of a glassy-white color, and semi-transparent. When present, they are generally found in considerable numbers, and they burrow many channels through the flesh of the apple, converting it into a spongy substance of a dull-yellowish color.

The change to a pupa takes place within the fruit. The pupa is about one-eighth of an inch long, somewhat sticky on the surface, of an elongated, oval form, pointed at one end, and rounded at the other; the head, thorax, and wing-cases are black; the abdomen is dull yellow.

The perfect insect very much resembles the Hessian fly in appearance, except that its legs are not so long and slender. The head, antennæ, and thorax are black; the abdomen dusky, almost black, with a pale-yellow band at each of the sutures; beneath it is yellow, with a dusky patch on the middle of each segment; the tip of the abdomen, ovipositor, and legs, are black. The wings are dull hyaline, tinged with a smoky hue, and about one-fourth longer than the body.

This insect has not thus far proved very destructive, and from its habits is scarcely likely to become so.

No. 62.—The Apple Fly.
Drosophila ampelophila Loew.

This is a two-winged fly, known as the vine-loving pomace fly, very similar in its habits to the apple midge, but it usually attacks the earlier varieties, showing a preference for such as are sweet. The larva (see *a*, Fig. 144)

FIG. 144.

a   *b*

generally enters the apple where it has been bored by the codling worm, or through the punctures made by the apple curculio, and sometimes through the calyx when the apple is quite sound. In August the fly (see Fig. 144, *b*) matures and deposits eggs for another brood, and successive generations follow until winter begins. The pupæ may be found during the winter in the bottoms of apple-barrels, and in this inactive state they remain until the following season. Usually several insects are found in the same apple, and sometimes the fruit is almost alive with them, when, being rapidly riddled with their borings, it speedily decays.

No. 63.—The Apple Thrips.

Phlæothrips mali Fitch.

This is a very small insect, about one-eighteenth of an inch long. It is slender, of a blackish-purple color, with narrow, silvery-white wings. Occasionally apples are found early in August, small and withered, with a cavity near their tip, about the size of a pea, and the surface of a blackened color, appearing as if the cavity had been gnawed out. Within this may usually be found one of these apple thrips, which had probably taken up its residence on the fruit while it was very small, and by frequent puncturing day after day the apple has become stunted in growth, and finally withered.

This insect has never yet proved very injurious; should it ever become so, it would be a difficult one to exterminate. Syringing thoroughly with tobacco-water or a solution of whale-oil soap would probably prove efficacious.

No. 64.—The Ash-gray Pinion.

Lithophane antennata (Walker.)

This insect is a moth, the larva of which has occasionally

FIG. 145.

been found boring into young apples and peaches during the month of June. Fig. 145 illustrates its mode of procedure.

The caterpillar is pale green, with cream-colored spots, and a broad, cream-colored band along the sides. When full grown, it leaves the fruit and works its way under the surface of the ground, where it forms a very thin, filmy, silken cocoon, within which it changes to a reddish-brown chrysalis.

The moth escapes in the autumn, and is of a dull ash-gray color, with its fore wings variegated with darker gray, or grayish brown, as shown in the figure.

SUPPLEMENTARY LIST OF INJURIOUS INSECTS WHICH AFFECT THE APPLE.

In addition to those already enumerated, the following insects are injurious to the apple, but, since they are more destructive to other fruits, they will be referred to under other headings.

ATTACKING THE BRANCHES.

The pear-blight beetle, No. 68; the New York weevil, No. 100; and the red-shouldered Sinoxylon, No. 130.

ATTACKING THE LEAVES.

The tarnished plant-bug, No. 71; the pear-tree leaf-miner, No. 74; grasshoppers, No. 80; the gray dagger-moth, No. 84; the waved Lagoa, No. 89; the blue-spangled peach-tree caterpillar, No. 102; the Io emperor-moth, No. 112; the Ursula butterfly, No. 116; the basket or bag-worm, No. 120; the white-lined Deilephila, No. 136; the rose-beetle, No. 151; and the smeared dagger, No. 194.

ATTACKING THE FRUIT.

The melancholy Cetonia, No. 82; and the plum curculio, No. 94.

INSECTS INJURIOUS TO THE PEAR.

ATTACKING THE TRUNK.

No. 65.—The Pear-tree Borer.

Ægeria pyri (Harris).

This is a whitish larva resembling that of the peach-tree borer, but much smaller, which feeds chiefly upon the inner layers of the bark of the pear-tree. Its presence may be detected from its habit of throwing out castings resembling fine sawdust, which are readily seen upon the bark of the tree. Before the larva changes to a chrysalis it eats a passage through the bark, leaving only the thinnest possible covering unbroken. Retiring towards the interior, it changes to a chrysalis, and late in the summer the chrysalis wriggles itself forward, and, pushing against the paper-like covering which conceals its place of retreat, ruptures it, and, projecting itself from the orifice, the moth soon bursts its prison-house and escapes, leaving nothing but the empty skin behind it.

The moth (Fig. 146) is somewhat like a small wasp, of a purplish or bluish-black color, with three golden-yellow stripes on its abdomen; the edges of the collar, the shoulder-covers, and the fan-shaped brush on the tail are of the same golden-yellow hue. The wings, which, when expanded, measure more than half an inch across, are clear and glass-like, with their veins and fringes purplish black, and across the tips of the fore wings is a broad dark band with a coppery lustre. The under side is pale yellow.

FIG. 146.

Remedies.—The trees should be examined in the spring, and if evidences of the presence of these larvæ are found, they should be searched for and destroyed. As a preventive measure, paint the trees with the mixture of soft-soap and

140

solution of soda, as recommended for the round-headed borer
of the apple (No. 2), or mound the trees about midsummer
with earth, as recommended for the peach-tree borer (No. 97).

No. 66.—The Pigeon Tremex.

Tremex Columba Linn.

The female Pigeon Tremex is represented in Fig. 147. It
is a large wasp-like creature, which measures, when its wings
are expanded, nearly two
inches across. The body is
cylindrical, and about an
inch and a half long ex-
clusive of its boring instrument, which projects about
three-eighths of an inch beyond the body. The wings
are of a smoky-brown color,
and semi-transparent; the
head and thorax are reddish,
varied with black, and the
abdomen is black, crossed by

FIG. 147.

seven yellow bands, all except the first two interrupted in
the middle. The horny tail and a round spot at its base are
ochre-yellow.

The male (Fig. 148) is unlike the female: it is smaller and
has no borer. Its wings are more transparent; the body is
reddish, varied with black, in form
somewhat flattened, rather wider behind, and ends with a conical horn.
The length of the body is from three-
fourths of an inch to an inch or more,
and the wings expand about an inch
and a half.

FIG. 148.

The female bores into the wood of
the tree with her borer, and, when the
hole is made deep enough, drops an egg into it. The egg is

oblong-oval, pointed at both ends, and rather less than one-twentieth of an inch in length.

The larva is soft, yellowish white, of a cylindrical form, rounded behind, with a conical horny point on the upper part of the hinder extremity, and when mature is about an inch and a half long. It bores deeply into the interior of the wood. Besides the pear, it is injurious to the buttonwood, elm, and maple.

From its secluded habits, this insect is a difficult one to cope with; fortunately, it is seldom present in sufficient numbers to be very injurious. It is said to be destroyed by Ichneumon flies, species of Pimpla, furnished with very long ovipositors, with which they bore into the trunks of trees inhabited by these Tremex larvæ, and deposit their eggs in them: these hatch into grubs, which consume their substance and cause their death.

ATTACKING THE BRANCHES.

No. 67.—The Twig-girdler.

Oncideres cingulatus (Say).

Fig. 149.

This beetle nearly amputates pear twigs during the latter half of August and the early part of September. The female makes perforations (Fig. 149, *b*) in the smaller branches of the tree upon which she lives, and in these deposits her eggs, one of which is shown of the natural size at *e*. She then proceeds to gnaw a groove about one-tenth of an inch wide and about a similar depth all around the branch, as shown in the figure, when the exterior portion dies, and the larva, when hatched, feeds upon the dead wood. The girdled twigs sooner or later fall to the ground, and in them the insect completes its transformations, and finally escapes as a perfect beetle. This insect is about eleven-twentieths of an inch in length, with a robust body of a brownish-gray color with dull reddish-yellow dots, and having a broad gray band across the middle of the wing-cases. The antennæ are longer than the body. The beetle is more common on the hickory than on the pear.

To subdue the insect, the dead and fallen twigs should be gathered and burnt.

No. 68.—The Pear-blight Beetle.
Xyleborus pyri (Peck).

During the heat of midsummer, twigs of the pear-tree sometimes become suddenly blighted, the leaves and fruit wither, and a discoloration of the bark takes place, followed by the speedy death of the part affected. Most frequently these effects are the result of fire-blight, a disease produced by a species of micrococcus, but occasionally they are due to the agency of the pear-blight beetle. In these latter instances there will be found, on examination, small perforations like pin-holes at the base of some of the buds, and from these issue small cylindrical beetles, shown magnified in Fig. 150, about one-tenth of an inch long, of a deep brown or black color, with antennæ and legs of a rusty red. The thorax is Fig. 150. short, very convex, rounded and roughened; the wing-covers are thickly but minutely punctated, the dots being arranged in rows; the hinder part of the body terminates in an abrupt and sudden slope.

The beetle deposits its eggs at the base of the bud, and when hatched the young larva follows the course of the eye of the bud towards the pith, around which it passes, consuming the tissues in its course, thus interfering with the circulation and causing the twig to wither. The larva changes to a pupa, and subsequently to a beetle, in the bottom of its burrow, and makes its escape from the tree in the latter part of June or the beginning of July, depositing its eggs before

August has passed. The hole made by the beetle when it is escaping is a little more than one-twentieth of an inch in diameter.

It was formerly supposed that these insects infested only such trees as were unhealthy or were already dying, but it has been stated that sound and healthy trees are attacked and severely injured by them. Neither are they limited in their operations to the twigs, but sometimes attack the trunk also. It is said that there are two broods each year, the early one nurtured in the trunk, and when these reach maturity, the newly-grown twigs, offering a more dainty repast, are subsequently invaded and destroyed.

The injuries inflicted by this insect are not confined wholly to the pear; occasionally it is found on the apple, apricot, and plum. The only remedy which has been suggested is to cut off the blighted limbs below the injured part and burn them before the beetle has escaped.

The damage caused by this insect must not be confounded with the well-known fire-blight on the pear, since that, as already remarked, is a disease of a totally different character, and is entirely independent of insect agency.

No. 69.—The Pear-tree Bark-louse.

Lecanium pyri (Schrank).

This insect is found on the under side of the limbs of young and thrifty pear-trees, adhering closely to the bark. It appears in the form of a hemispherical scale about one-fifth of an inch in diameter, of a chestnut-brown color, sometimes marked with faint blackish streaks, and having on its surface some shallow indentations. The outer margin is wrinkled. These scales, when mature, are the dead bodies of the females covering and protecting their young ; some are darker in color than others, and there are some smaller ones which are of a dull-yellow hue.

Under the scales the young lice are interspersed through a mass of white cotton-like matter, which subsequently increases in volume and protrudes from under the scale. Early in the season they crawl out and distribute themselves over the smooth bark, appearing as minute whitish specks. When magnified, they are found to be of an oval form, somewhat flattened, about one-hundredth of an inch long, of a dull-white color, with six legs and short antennæ. The young lice attach themselves to the bark, which they puncture with their beaks, living on the sap, and during the season materially increase in size. They pass the winter in a torpid state, and in the spring the males enter the pupal condition, and subsequently appear as minute two-winged flies, while the females gradually grow to the size and form of the scales referred to, and after depositing their eggs die, when their dried bodies remain to serve as a shelter for their offspring. This is believed to be identical with the bark louse which occurs upon the pear-tree in Europe.

Remedies.—Fortunately, these insects are of such a size that they are easily seen. They should be looked for during the latter part of June, at which time the females will have attained their full size, and, when discovered, should be promptly removed. The under side of the limbs should also be well scrubbed with a brush dipped in some alkaline solution.

A small, four-winged parasite lives in the bodies of the females, feeds upon their substance and destroys them, and forms a chrysalis under the scale. When this fly matures, it gnaws a round hole through the scale and escapes.

No. 70.—The Pear-tree Psylla.

Psylla pyri Schmidb.

During the middle of May, when growth is rapid, the smaller limbs and twigs of pear-trees are sometimes observed to droop; a close examination reveals a copious exudation of sap from about the axils of the leaves, so abundant that it drops upon the foliage below, and sometimes runs down the branches to the ground. Flies and ants gather around in crowds to sip

10

the sweets, and by their busy bustle draw attention to the mischief progressing. With a magnifying lens the authors of the injury may be observed immersed in the sap about the axils of the leaves.

This insect is known as the Pear-tree Psylla, a small, yellow, jumping creature, flattened in form, and provided with short legs, a broad head, and sharp beak. With the beak are made the punctures from which the sap exudes. In rare instances they occur in immense

Fig. 151.

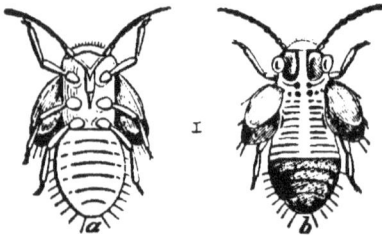

numbers, when almost every leaf on a tree will seem to be affected; all growth is at once arrested, and frequently the tree loses a considerable portion of its leaves. When in the pupa state with the wings developing, they present the appearance shown in Fig. 151; *a* represents the under side, *b* the upper side; the perfect winged insect is shown in Fig. 152, all highly magnified.

Fig. 152.

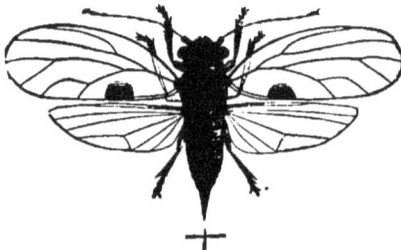

The color of the pupa is deep orange-red, the thorax striped with black, and the abdomen blackish brown.

Towards the end of the summer they attain maturity, when they are furnished with transparent wings; the head is deeply notched in front; color orange-yellow, with the abdomen greenish. Length one-tenth of an inch.

Remedies.—Paint the twigs with a strong solution of soft-soap, as recommended for No. 2, or syringe the trees with strong soapsuds.

ATTACKING THE BUDS.

No. 71.—The Tarnished Plant-bug.

Lygus lineolaris (P. Beauv.).

This insect, which is represented magnified in Fig. 153, is about one-fifth of an inch long, and varies in color from dull dark brown to a greenish or dirty yellowish brown, the males being generally darker than the females. The head is yellowish, with three narrow, reddish stripes; the beak or sucker is about one-third the length of the body, and when not in use is folded upon the breast. The thorax has a yellow margin and several yellowish lines running lengthwise; behind the thorax is a yellow V-like mark, sometimes more or less indistinct. The wings are dusky brown, and the legs dull yellow.

Fig. 153.

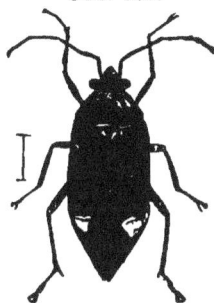

It passes the winter in the perfect state, taking shelter among rubbish, or in other convenient hiding-places, and early in May, as soon as vegetation starts, it begins its depredations. Concealing itself within the young leaves of the expanding buds of the pear, it punctures them about their base and along their edges, extracting their juices with its beak. The puncture of the insect seems to have a poisonous effect, and the result is to disfigure and sometimes entirely destroy the young leaves, causing them to blacken and wither. These insects are also partial to the unopened buds, piercing them from the outside, and sucking them nearly dry, when they also become withered and blackened. Sometimes a whole branch will be thus affected, being first stunted, then withering, and finally dying. Early in the morning these plant-bugs are in a sluggish condition, and may be found buried in the expanding leaves, but as the day advances and the temperature rises they become active, and when approached dodge quickly about from place to place, drop to the ground, or else take wing and fly away. In common with most true bugs, they have when handled a disagreeable odor. In the course of two or three weeks they disappear, or cease to be sufficiently injurious to attract attention.

It is stated that they deposit their eggs on the leaves, and that later in the season the young and old bugs may be found together. The young bugs are green, but in other respects do not differ from their parents, except in lacking wings. While they seem particularly partial to the pear, they attack also the young leaves of the quince, apple, plum, cherry, and strawberry, as well as those of many herbaceous plants.

Remedies.—First of all, clean culture, so as to leave no shelter for the bug in which to winter over. When they appear in spring, shake them from the trees very early in the morning, while they are in a torpid state, and destroy them.

No. 72.—The Oak Platycerus.

Platycerus quercus (Weber).

This is an insect belonging to the family of stag beetles, which has occasionally been found injurious to pear-trees in Illinois by devouring the buds. In the larval state it feeds on decaying wood in old oak logs and stumps. It matures and appears as a beetle about the time that the buds of the pear are bursting, and continues feeding for many days, completely eating out the swelling buds and the ends of the new shoots.

Fig. 154.

It is a blackish beetle, of a greenish cast, with ribbed wing-covers, and nearly half an inch in length. It is represented in Fig. 154. As this has hitherto been comparatively a rare beetle, it is scarcely likely ever to prove generally troublesome to pear-growers.

ATTACKING THE FLOWERS.

No. 73.—The Pear-tree Blister-beetle.

Pomphopœa aenea (Say).

This is a greenish-blue or brassy-looking beetle, rather more than half an inch long (see Fig. 155), with head and thorax punctated and somewhat hairy, the wing-cases roughened and with two slightly-elevated lines.

Fɪɢ. 155.

These beetles have been found injurious to pear-blossoms both in Michigan and in Pennsylvania. They begin their work by devouring the corolla, then the pistil and calyx, and a portion of the forming fruit, but are said to avoid the stamens. They will occasionally eat small portions of the tender foliage, and are usually most abundant on the tops of the trees and about the extremities of the limbs. They also attack the blossoms of the cherry, plum, and quince, but have not been observed on the apple or peach.

This pest is easily controlled. On jarring the trees they drop at once to the ground, and if taken in the cool of the morning are very sluggish in their movements. Later in the day, in the heat of the sun, they become much more active, and fly readily.

ATTACKING THE LEAVES.

No. 74.—The Pear-tree Leaf-miner.

Lithocolletis geminatella Packard.

The larva of this insect mines the leaves of the pear, and also those of the apple. It is very small, of a pale-reddish color, with a black head and a black patch on the upper part of the next segment. In Fig. 156 it is shown magnified. It usually draws two leaves together and fastens them with silken fibres, or else folds one up and eats the surface, making unsightly blotches, which disfigure and injure the leaves. About the middle of August, the larva changes to a long, slender chrysalis within this mine (Fig. 157, also magnified). The moths appear a few days afterwards.

Fig. 156.

Fig. 157.

When its wings are expanded, the moth (Fig. 158, enlarged) measures about one-third of an inch across. The fore wings are dark gray, with a round blackish spot on the middle of the inner edge of the wing, which is not shown in the figure, also an eye-like spot on the outer edge, with a black pupil.

Fig. 158.

As the season advances, these insects sometimes become very abundant, and towards the end of autumn a large proportion of the leaves of the pear and apple trees become blotched and disfigured from their work. Since they pass the winter in the larval or chrysalis condition in their leafy enclosures, their numbers may be materially reduced by gathering all the fallen leaves in the autumn and burning them.

No. 75.—The Pear-tree Slug.

Selandria cerasi Peck.

In the year 1790, Prof. Peck, of Massachusetts, wrote a pamphlet entitled "Natural History of the Slug-worm," which was printed in Boston the same year by order of the Massachusetts Agricultural Society and was awarded the Society's premium of fifty dollars and a gold medal. Although more than ninety years have passed since that pamphlet was written, not much has been added in the interval to our knowledge of the history and habits of this insect. In the mean time, however, it has spread over the greater portion of the United States and Canada, injuring more or less seriously the foliage of our pear, cherry, quince, and plum trees every year.

This insect passes the winter in the pupa state under ground; the flies, the progenitors of the mischievous brood of slugs, appearing on the wing in the Northern States and Canada from about the third week in May until the middle of June. The fly (Fig. 159) is of a glossy black color, with four transparent wings, the front pair being crossed by a dusky cloud; the veins are brownish, and the legs dull yellow, with black thighs, except the hind pair, which are black at both extremities, and dull yellow in the middle. The female fly is more than one-fifth of an inch long; the male is somewhat smaller. When the trees on which these flies are at work are jarred or shaken, or if the flies are otherwise disturbed, they fall to the ground, where, folding their antennæ under their bodies and bending the head forward and under, they remain for a time motionless.

Fig. 159.

The saw-flies have been so called from the fact that in most of the species the females are provided with a saw-like appendage at the end of the body, by which slits are cut in the leaves of the trees, shrubs, or plants on which the larvæ feed, in which slits the eggs are deposited. The female of this species begins to deposit her eggs early in June; they are placed singly within little semicircular incisions through the skin of the leaf, sometimes on the under side and sometimes on the upper. In about a fortnight these eggs hatch.

The newly-hatched slug is at first white, but soon a slimy matter oozes out of the skin and covers the upper part of the body with an olive-colored sticky coating. After changing its skin four times, it attains the length of half an inch or more (see Fig. 160, *a*), and is then nearly full grown. It is a disgusting-looking creature, a slimy, blackish, or olive-brown slug, with the anterior part of its body so swollen as to re-

semble somewhat a tadpole in form, and having a disagreeable and sickening odor. The head is small, of a reddish color, and is almost entirely concealed under the front segments. It is of a dull-yellowish color beneath, with twenty very short legs, one pair under each segment except the fourth and the last. After the last moult it loses its slimy appearance and dark color, and appears in a clean yellow skin entirely free from slime ; its form is also changed, being proportionately longer. In a few hours after this change it leaves the tree and crawls or falls to the ground, where it buries itself to a depth of from one to three or four inches. By repeated movements of the body the earth is pressed firmly on all sides, and an oblong-oval chamber is formed, which is afterwards lined with a sticky, glossy substance, which makes it retain its shape. Within this little earthen cell the insect changes to a chrysalis, and in about a fortnight finishes its transformations, breaks open the en-closure, crawls to the surface of the ground, and appears in the winged form.

Fig. 160.

About the third week in July the flies are actively engaged in depositing eggs for a second brood, the young slugs appear-ing early in August. They reach maturity in about four weeks, then retire under ground, change to pupæ, and remain in that condition until the following spring.

Pear and cherry growers should be on the lookout for this destructive pest about the middle of June, and again early in August, and if the young larvæ are then abundant they should be promptly attended to, since if neglected they soon play sad havoc with the foliage, feeding upon the upper side of the leaves and consuming the tissues, leaving only the veins and under skin. The foliage, deprived of its substance, withers and becomes dark-colored, as if scorched by fire, and soon after-wards it drops from the trees. In a badly-infested pear orchard, whole rows of trees may sometimes be seen as bare of foliage during the early days of July as they are in midwinter. In such instances the trees are obliged to throw out new leaves; and this extra effort so exhausts their vigor as to interfere seriously with their fruit-producing power the following year. Although very abundant in a given locality one season, these slugs may be very scarce the next, as they are liable to be destroyed in the interval by enemies and by unfavorable climatic influences.

Remedies.—Hellebore in powder, mixed with water in the proportion of an ounce to two gallons, and applied to the foliage with a syringe or a watering-pot, promptly destroys this slug; and Paris-green, applied in the same manner, in the proportion of a teaspoonful to the same quantity of water, would doubtless serve a similar purpose. Fresh air-slaked lime dusted on the foliage is said to be an efficient remedy. It has been recommended to dust the foliage with sand, ashes, and road dust, but these are unsatisfactory measures, and of little value. A very minute Ichneumon fly is said to lay its eggs within the eggs of this saw-fly, and from its tiny egg a little maggot is hatched, which lives within the egg of the saw-fly and consumes it.

No. 76.—The Green Pear-tree Slug.

Another species of saw-fly, as yet undetermined, also attacks the leaves of the pear. The larvæ appear from about the first to the middle of June, and eat holes in the leaves or semi-circular portions from the edge. They are about half an inch in length, nearly cylindrical in form, tapering slightly towards the hinder segments. The head is rather small, pale green with a yellowish tinge, and has a dark-brown dot on each side; the jaws are tipped with brown. The body above is semi-transparent, of a grass-green color faintly tinged with yellow, the yellow most apparent on the posterior segments; there is a line down the back of a slightly deeper shade of green, and one along each side, close to the under surface, of a paler hue.

The under side is similar to the upper; feet whitish green, semi-transparent.

About the middle of June this larva seeks some suitable hiding-place, such as a crevice in the bark of the tree, or other shelter, and there makes and fastens firmly a small, brownish, papery-looking cocoon, in which it undergoes its transformations and remains until the following spring, when the perfect fly appears.

The fly bears a general resemblance to that of the pear-tree slug, but is smaller.

The remedies applicable to the pear-tree slug would serve equally well in this instance; but these insects are seldom found in sufficient abundance to require a remedy.

No. 77.—The Goldsmith-beetle.

Cotalpa lanigera (Linn.).

This is, without doubt, one of the most beautiful of all our leaf-eating beetles. It is nearly an inch in length (see Fig. 161), of a broad, oval form, with the wing-cases of a rich yellow color and pale metallic lustre, while the top of the head and the thorax gleam with burnished gold of a brilliant reddish cast. The under surface has a polished coppery hue, and is thickly covered with whitish, woolly hairs: this latter characteristic has suggested its specific name, *lanigera*, or wool-bearer.

Fig. 161.

This insect appears late in May and during the month of June, and is distributed over a very wide area, being found in most of the Northern United States and in Canada; and, although seldom very abundant, rarely does a season pass without some of them being seen. During the day they are inactive, and may be found clinging to the under side of the leaves of trees, often drawing together two or three leaves and holding them with their sharp claws for the purpose of concealing themselves. At dusk they issue from their hiding-places and fly about with a buzzing sound among the branches of trees, the tender leaves of which they devour. The pear, oak, poplar, hickory, silver abele, and sweet-gum all suffer more or less from their attacks. Like the common May-bug, this beautiful creature is attracted by light, and often flies into lighted rooms on summer evenings, dashing against everything it meets with, to the great alarm of nervous inmates. In some seasons they are comparatively common, and may then be readily captured by shaking the trees on which they are lodged, in the daytime, when they do not attempt to fly, but fall at once to the ground.

The beetle is short-lived. The female deposits her eggs in the ground at varying depths during the latter part of June, and, having thus provided for the continuance of her species, dies. The lives of the males are of still shorter duration. The eggs are laid during the night, the whole number probably not exceeding twenty; they are very large for the size of the beetle, being nearly one-tenth of an inch in length, of a long, ovoid form, and a white, translucent appearance.

In about three weeks the young larva is hatched; it is of a dull-white color, with a polished, horny head of a yellowish brown, feet of the same hue, and the extremity of the abdomen lead-color. The mature larva (Fig. 162) is a thick, whitish, fleshy grub, very similar in appearance to that of the May-bug, which is familiarly known as "the white grub." It lives in the ground and feeds on the roots of plants, and is thus sometimes very destructive to strawberry-plants. It is said that the larva is three years in reaching its full growth; finally, it matures in the autumn, and late the same season or early in the following spring changes to a beetle.

FIG. 162.

No. 78.—The Iridescent Serica.

Serica iricolor Say.

This beetle is said to have proved very injurious to pear-trees in New Jersey by devouring the leaves. It is of an oval form, about one-fifth of an inch long, of a dull bluish-black color, and clothed with long, fine, silky hairs, especially on the thorax; it is represented in Fig. 163. This insect has the same habit of dropping to the ground when the trees are jarred or shaken as the goldsmith-beetle (No. 77), and if it proves at any time troublesome it may be collected in this way and destroyed. It is not known how or where the larva of this species lives, but it probably dwells under ground and feeds on the roots of plants.

Fig. 163.

No. 79.—The Pear-tree Aphis.

An undetermined species of aphis sometimes attacks the leaves of the pear-tree early in June, causing them to twist and curl up very much. In the pupa state these insects are active, with the wings partly developed. They are then green, with a row of brownish dots along the back, which are smaller on the anterior segments and larger on the middle ones; there are also some streaks of the same color along each side. The wings are enclosed in cases on the sides about half the length of the body; body plump; honey-tubes pale whitish, tipped with black; feet pale whitish. All the specimens seen at this time have partly or fully developed wings.

In the perfect winged specimens the head is black; thorax black above, greenish below; body brownish black above, green on the sides and beneath, with a few blackish dots; antennæ brownish black. When the insect escapes from the pupa state, the empty pupa skin is left attached to the under surface of the curled leaves.

The remedies recommended for the apple-tree aphis (No. 57) will be serviceable for this insect also.

No. 80.—Grasshoppers, or Locusts.

In addition to the insects already treated of, several species of grasshoppers, or, more correctly, locusts, attack the leaves of the pear, and, when abundant, will often entirely strip young trees of their foliage. In Fig. 164 we have a representation of the red-legged locust, *Caloptenus femur-rubrum* (De Geer), one of our commonest species, which is abundant everywhere, from Maine to Minnesota, throughout the greater portion of Canada, and from Pennsylvania to Kansas. In Fig. 165 is shown the noted Rocky Mountain locust, *Calop-*

Fig. 164.

Fig. 165.

tenus spretus Thomas, which has proved so terribly destructive in the West and Northwest. Although much resembling the red-legged locust in size and general appearance, the wings are longer, and there are other points of difference which enable the entomologist readily to separate the species. These, however, need not be enumerated here. In Fig. 166 the females of the Rocky Mountain locust are depicted at *a, a, a,* in the act of depositing their eggs. These eggs are laid in the ground in masses, in which the eggs are carefully arranged, and the whole coated with a gummy covering. In the lower part of the figure one of the egg-masses is shown with one end open, others in position at *d* and *e,* and the eggs separated at *c; f* shows where an egg-mass has been deposited and the aperture closed.

In Fig. 167 another common species is represented,—at *a* in the immature or larval state, at *b* in the mature or perfect condition. This insect is known under the name of the green-faced locust, *Tragocephala viridifasciata* (De Geer). There are many other species which might be referred to, but

FIG. 166.

these will suffice to illustrate the family, all the members of which are destructive, especially during the latter part of the summer.

When young trees are deprived of their leaves in the midst of their growth, they fail to ripen their wood properly, and their vitality is weakened so that they are more liable to

FIG. 167.

injury from winter, and also more prone to disease. Grasshoppers do not confine their attacks to the pear, but devour also the leaves of young apple, plum, and other trees.

To destroy these pests, the trees, when not fruiting, may be syringed with Paris-green and water in the proportion of two teaspoonfuls of the poison to two gallons of water.

ATTACKING THE FRUIT.

No. 81.—The Indian Cetonia.

Euphoria Inda (Linn.).

This is one of the earliest insect visitors in spring, appearing towards the end of April or in the beginning of May, when it flies about in dry fields on the borders of woods on sunny days, making a loud buzzing sound like a bee. It is little more than half an inch in length (see Fig. 168), and has a broad body, obtuse behind. The head and thorax are of a blackish copper-brown, thickly covered with short, greenish-yellow hairs. The wing-cases are light yellowish brown, with a number of irregular black spots. The under side of the body is black and very hairy; the legs are dull red. A variety of this species is occasionally met with entirely black.

FIG. 168.

The early brood are fond of sucking the sweet sap which exudes from wounded trees or freshly-cut stumps; in September a second brood appear, and these injure fruits, burrowing into ripe pears almost to their middle, revelling on their sweets, and inducing rapid decay. They also attack peaches and grapes.

Nothing has yet been recorded in reference to the larval history of this species. It is probable that the late brood of beetles hibernate, passing the winter in a torpid state, hidden in sheltered places, and awakening with the return of spring, when they issue from their retreats, after which, having deposited eggs for another brood, they die.

The only remedy suggested for these insects is to catch and destroy them. They are seldom very abundant.

No. 82.—The Melancholy Cetonia.

Euphoria melancholica (Gory).

This insect belongs to the same genus as the Indian Cetonia (No. 81), and is very similar to it in appearance and habits, but is somewhat smaller. (See Fig. 169.)

Fig. 169.

This beetle has also been found eating into ripe pears, and occasionally apples. It is found in the South in cotton-bolls, in the holes left by the boll-worm. It appears to frequent the bolls for the purpose of consuming the exuding sap.

SUPPLEMENTARY LIST OF INJURIOUS INSECTS WHICH AFFECT THE PEAR.

ATTACKING THE ROOT.

The broad-necked Prionus, No. 122, is occasionally very destructive to the roots of the pear.

ATTACKING THE TRUNK.

The round-headed apple-tree-borer, No. 2, and the flat-headed apple-tree borer, No. 3, both injure the pear, and are often found under the bark, especially about the base of the trunk.

ATTACKING THE BRANCHES.

The apple-twig borer, No. 13 ; the oyster-shell bark-louse, No. 16 ; the scurfy bark-louse, No. 17 ; and the New York weevil, No. 100, all affect the branches of the pear-tree.

ATTACKING THE LEAVES.

Many of the insects which devour the leaves of other fruit-trees feed also on those of the pear, such as the white-marked tussock-moth, No. 22 ; the red-humped apple-tree caterpillar, No. 24; the fall web-worm, No. 27; the Cecropia emperor-moth, No. 28; the oblique-banded leaf-roller, No. 35; the eye-spotted bud-moth, No. 38; the blue-spangled peach-tree caterpillar, No. 102; and the basket-worm, or bag-worm, No. 120.

ATTACKING THE FRUIT.

The codling moth, No. 58, so destructive to the fruit of the apple, is almost equally injurious to that of the pear. The plum curculio, No. 94, and the quince curculio, No. 121, also affect this fruit.

11

ATTACKING THE LEAVES.

No. 83.—The Plum-tree Sphinx.

Sphinx drupiferarum (Sm. & Abb.).

The moths belonging to the family known as Sphinx moths are peculiar in their form and habits. Their bodies are robust, and their wings are usually long and narrow and possess great strength and capacity for rapid flight. On the wing they much resemble humming-birds, and hence are frequently called

FIG. 170.

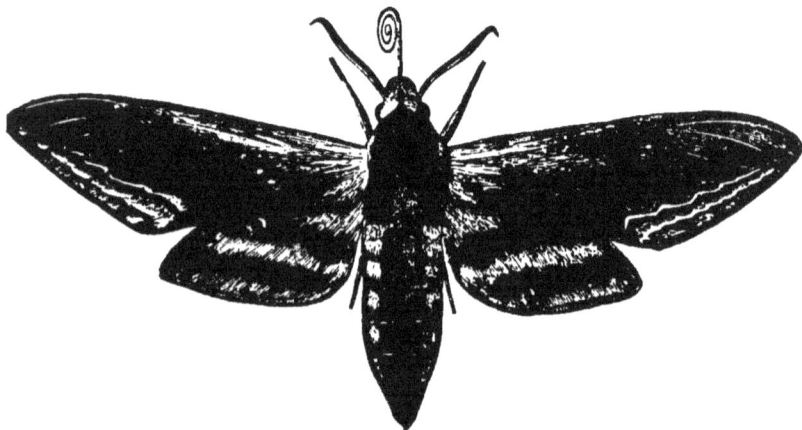

humming-bird moths. Most of the species remain torpid during the day, but become active about dusk, when they may be seen poising in the air over some flower, with their wings rapidly vibrating, and producing a humming sound.

The plum sphinx is a handsome insect, and is well represented in Fig. 170. It appears as a moth during the month of June; its body is about an inch and a half long, and its

162

wings expand from three and a half to four inches. The wings are of a purplish-brown color, the anterior pair having a stripe of white on their front edge, and one of a fawn color on their outer edge; there are also three or four oblique black streaks, and a black dot on the white stripe. The hind wings have two whitish, wavy stripes, with a fawn-colored stripe also on their outer edge. The head and thorax are blackish brown, with a whitish-fawn color at the sides; the eyes are very prominent, and the snout-like projection in front consists of the two palpi or feelers, within which lies the proboscis or tongue, snugly coiled up between them like the mainspring of a watch; in the figure this proboscis is shown partly extended. When stretched to its full length, it is as long as the body, and is used by the insect in extracting honey from flowers. The body is brown, with a central line and a band on either side of black, the latter containing four or five dingy-white spots.

The moth deposits her eggs singly on the leaves of the plum. The egg is about one-fifteenth of an inch long, slightly oval, with a smooth surface, and of a pale yellowish-green color. It hatches in from six to eight days, when the young larva eats its way out through the side of the egg; its first meal is usually made from the egg-shell, which it partly or wholly devours.

The newly-hatched larva is one-fourth of an inch long, of a pale yellowish-green color, with a few slightly-elevated whitish tubercles on every segment, from each of which arises a single fine short hair; the caudal horn is black. The full-grown caterpillar is about three and a half inches long (see Fig. 171), of a beautiful apple-green color, with a lateral dark-brown or blackish stripe. On each side of the body there are seven broad oblique white bands, bordered in front with light purple or mauve; the stigmata or breathing-pores, which are ranged along each side of the body, are of a bright orange-yellow. The caudal horn is long, dark brown, with a yellowish tint about the base at the sides. After satisfying its rapacious appetite, this larva often assumes for a time the peculiar rigid appearance shown in the cut. Though presenting a formidable aspect, it is perfectly harmless, and may

Fig. 171.

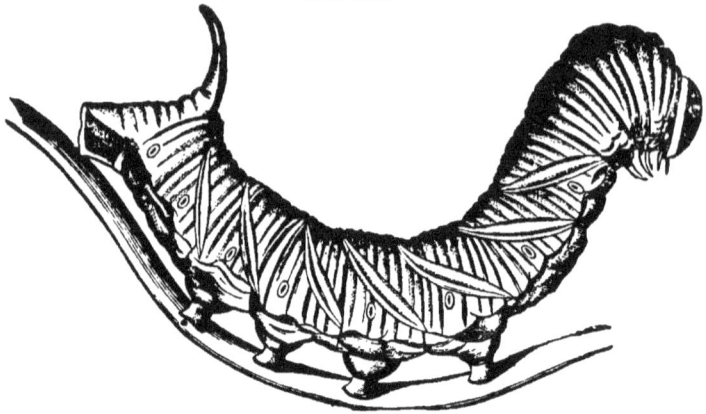

be handled with impunity; it may be found on the trees from the middle of July to the end of August.

When mature, the caterpillar descends to the ground, and, having buried itself under the surface to the depth of several inches, prepares a convenient chamber, which it lines with a gummy, water-proof cement, and there changes to a chrysalis, as shown in Fig. 172, which is about an inch and a half long,

Fig. 172.

of a dark reddish-brown color, with a short, thick, projecting tongue-case. The insect remains in the ground in this condition until the following June; indeed, occasionally specimens have been known to remain in this torpid state until the spring of the second year following.

The ravages of the plum-tree sphinx are never very extensive, yet it appears at times in some localities in sufficient numbers to cause annoyance. The denuded twigs promptly attract the attention of the vigilant fruit-grower, who will soon search out and exterminate the destroyer.

No. 84.—The Gray Dagger-moth.

Apatela occidentalis (G. & R.).

This is a pretty, pale, silvery-gray moth, the first brood of
which appear on the wing late in May or early in June. It
is shown in Fig. 173. The fore wings are pale gray, with
various black lines or markings,
the principal one being in the
form of an irregular cross, bearing
a resemblance to the Greek letter
F placed sideways; this is situated about the middle of the fore
wing, towards the outer edge. A

Fig. 173.

second smaller mark of the same character is found between
this and the tip of the wing; a black line proceeds from the
base of the wing and extends to near the middle. The hind
wings are dark glossy gray; the edges of both pairs have a
whitish fringe, with an inner border of black spots; the body
is gray. The wings, when expanded, measure from an inch
and a half to two inches across.

The moths deposit their eggs singly on the leaves of plum,
cherry, and apple trees, and the caterpillar becomes full
grown during the first or second week in July. It is then
about an inch and a half long. Its head is rather large,
flat in front, black, with yellowish dots at the sides. The
body is bluish gray above, with a wide slate-colored band
down the back, in which is a central pale-orange line from
the second to the fifth segment. From the fifth to the
eleventh, inclusive, each segment is ornamented with a beautiful group of spots, placed in the dorsal band, two of them
bright orange, one in front and one behind, and one of a
greenish metallic hue on each side, each group being set in a
nearly circular patch of velvety black. There are two cream-
colored stripes on the sides, which become indistinct towards
each extremity, and into which there extends from each of the
black dorsal patches a short, black, curved line, having behind

its base a yellowish dot; the sides are marked with dull ochrey spots, and on the top of the twelfth segment there is a prominent black hump. The body is sparingly covered with whitish hairs, which are distributed chiefly along the sides. The under surface is of a dull-greenish color; the feet are black.

When full grown, this larva spins a slight cocoon in some sheltered spot, and there changes to a chrysalis, about seven-tenths of an inch long, of a reddish-brown color, with a polished surface. From these the second brood of moths appear late in July, and shortly after eggs are again deposited, from which the later brood of larvæ mature about the middle of September, which then become chrysalids, and produce moths the following spring.

This insect seldom occurs in sufficient numbers to prove very destructive; should it ever do so, it may be readily destroyed by syringing the trees with powdered hellebore or Paris-green mixed with water, as recommended for the pear-tree slug (No. 75). The larvæ are often captured under the bands set as traps for the larvæ of the codling moth.

No. 85.—The Mottled Plum-tree Moth.

Apatela superans (Guen.).

The caterpillar of this moth also feeds on the leaves of the plum, and, like that last described, is solitary in its habits. It appears about the middle of June. It is a green caterpillar, about an inch long, with its body seeming as if laterally compressed, making it appear higher than it is wide. There is a

FIG. 174.

broad chestnut-colored stripe along the back, margined with yellowish, and on every segment there are several shining tubercles, each giving rise to one or more blackish hairs; there are also a few whitish hairs along the sides of the body. Fig. 174 represents a partly-grown specimen of this or a very closely allied species.

About the middle of July the moth (Fig. 175) escapes from the cocoon. The thorax and abdomen are gray, dotted with black points; fore wings gray, with black or brownish-black markings; hind wings brownish gray. When ex-panded, the wings measure about an inch and a half across.

FIG. 175.

This species is double-brooded. The moths that appear in July deposit eggs from which hatch larvæ which reach maturity in September, enter the chrysalis state, and remain in this condition until the following spring. An Ichneumon fly attacks this species and destroys many of them. They are seldom numerous, and never likely to prove very troublesome.

No. 86.—The Horned Span-worm.

Nematocampa filamentaria Guen.

This singular-looking caterpillar is frequently found on plum-trees, devouring the leaves; it is also found on maple, oak, and probably other trees, and on strawberry-vines. It is about seven-tenths of an inch long (see Fig. 176), of a grayish color, with dusky and blackish streaks. On the hinder part of the fifth segment are two long, curved, fleshy horns extending forward, and on the sixth segment there is a similar pair curving backwards. The head is spotted with brown. There are two short brown tubercles on the posterior part of the fourth segment, and two small gray warts on each of the segments behind, those on the eleventh being most prominent. It may be found during the first half of June, and sometimes later. During the latter part of the month it constructs a slight cocoon composed of pieces of leaves fastened together with silken threads, and within this enclosure changes to a reddish-gray or pale-brown

FIG. 176.

chrysalis, in which state it continues about ten days, when the
perfect insect escapes.

This is a small moth (Fig. 177), which measures, when its
wings are spread, from three-quarters of an inch to an inch
across.   It is of a pale ochreous color, with

Fig. 177.

reddish-brown lines and dots, a ring on the
discal space, and just beyond it a dark, lead-
colored band, which becomes an almost square
patch on the inner angle and is continuous
with a broad band of the same color on the
hind wings.   The moths are on the wing in July and early
in August.   This is never likely to become a very injurious insect, but, from its unique appearance, it will always
attract attention.

No. 87.—The Disippus Butterfly.

Limenitis disippus Godt.

This is one of our common butterflies, the larva of which
is occasionally found feeding on the leaves of plum-trees.

Fig. 178.

The wings of the butterfly are of a warm orange-red color,
with heavy black veins, and a black border with white spots.
In Fig. 178 the left wings represent the upper surface, while
those of the right, which are slightly detached from the body,
show the under side.   It appears on the wing during the

latter half of June and in July, and deposits its eggs, sometimes on the plum, but more frequently on the willow and
poplar.

The egg is less than one-twenty-fifth of an inch in length,
globular in form, and beautifully reticulated, as shown in Fig.

FIG. 179.

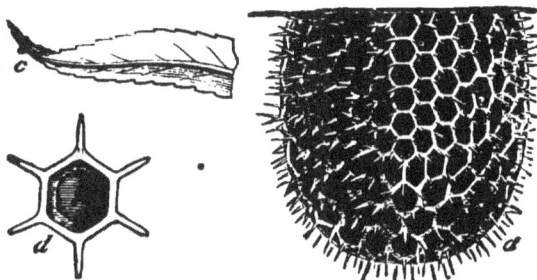

179, where *a* represents the egg highly magnified.   It is covered with short, transparent, hair-like spines.   One of the
hexagonal indentations, with its projecting filaments, is shown,
much enlarged, at *d*.   At first it is pale yellow, but as the

FIG. 180.

larva within develops it becomes pale gray ; the egg is generally laid on the under side of a leaf, near the tip, as seen
at *c* in the figure.   In a few days it hatches, and in about a
month the larva attains its full growth, when it presents the
appearance shown in Fig. 180, at *a*.

It is about an inch and a half in length; the head is pale

green, with two dull-white lines down the front, roughened
with a number of small green and greenish-white tubercles,
and tipped with two of a green color. The body above is
a rich dark green, with patches and streaks of creamy white;
the second segment is smaller than the head, and its surface
covered with many whitish tubercles; the third, dull whitish
green, raised considerably above the second, with a flat ridge,
having a long, brownish horn on each side, which is thickly
covered with very short spines. The fourth segment is similar
in size to the third, with the same sort of ridge above, and a
small tubercle on each side, tipped with a cluster of short,
whitish spines. On each segment behind these there are two
tubercles emitting clusters of whitish spines, those on the sixth
and twelfth being much larger than the others, while on each
segment behind the fourth, except the ninth, there are several smaller tubercles of a blue color. There are two large
patches of white on the upper part of the body, and a band
of the same color along each side.

When about to change to a chrysalis, the caterpillar suspends
itself, head downwards, and, shedding its skin, appears as at *b*,
Fig. 180, and in about ten or twelve days the butterfly escapes.

There are two broods of this insect during the year. The
larvæ from the eggs deposited by the second brood of butterflies hibernate when less than half grown, and complete their
growth the following spring. They construct from part of
the leaf a curious little case,
shown at *c*, in Fig. 180, which,
being firmly fastened to the
branch by silken threads,
serves during the winter
months as a shelter and a
hiding-place. There are several parasites which reduce
the numbers of this insect;
one is a tiny, four-winged fly, which infests the eggs (*Tri-
chogramma minuta* Riley Fig. 181, where *a* represents the fly;

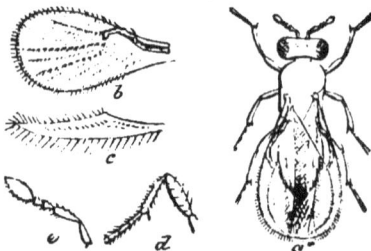

Fig. 181.

5, c, its fringed wings; d, one of its legs, and e, one of its antennæ). Another parasite is a small, black, four-winged fly, and a third a larger two winged-fly; the two latter attack the insect in its caterpillar state.

No. 88.—The Polyphemus Moth.
Telea polyphemus (Linn.).

The caterpillar of this insect, which is often found feeding on the leaves of plum-trees, is also known as the American silk-worm, in consequence of its having been extensively reared for the sake of its silk. When full grown, the larva presents the appearance shown in Fig. 182, and is over three

FIG. 182.

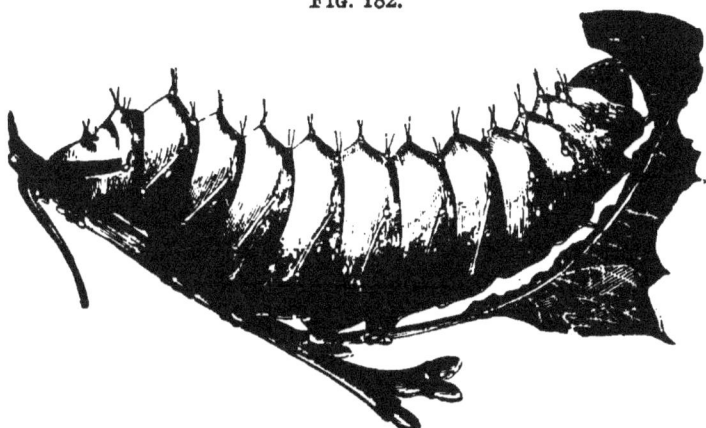

inches in length, with a very thick body. It is of a handsome light yellowish-green color, with seven oblique pale-yellowish lines on each side of the body; the segments, which have the spaces between them deeply indented, are each adorned with six tubercles, which are sometimes tinted with orange, have a small silvery spot on the middle, and a few hairs arising from each. The head and anterior feet are pale brown, the spiracles pale orange, and the terminal segment bordered by an angular band resembling the letter V, of a purplish-brown color.

When mature, the caterpillar proceeds to spin its cocoon within an enclosure usually formed by drawing together some of the leaves of the tree it has fed upon, some of which are firmly fastened to the exterior of the structure. The cocoon (Fig.

FIG. 183.

183) is a tough, pod-like enclosure, nearly oval in form, and of a brownish-white color, and within it the larva changes to an oval chrysalis, of a chestnut-brown color, represented in Fig. 184. Usually, the cocoons drop to the ground with the fall of the leaves, remaining there during the winter.

FIG. 184.

Late in May or early in June the prisoner escapes from its cell as a large and most beautiful moth, the male of which is shown in Fig. 185, the female in Fig. 186. The antennæ are feathered in both sexes, but more widely so in the male than in the female. The wings, which measure, when expanded, from five to six inches across, are of a rich buff or ochre-yellow color, sometimes inclining to a pale-gray or cream color, and sometimes assuming a deeper, almost brown shade. Towards the base of the wings they are crossed by an irregular pale-white band, margined with red; near the outer margin is a stripe of pale purplish white, bordered within by one of deep, rich brown, and about the middle of each wing is a transparent eye-like spot, with a slender line across its centre; those on the front wings are largest, nearly round, margined with yellow, and edged outside with black. On the hinder wings the spots are more eye-like in shape, are bordered with yellow, with a line of black edged with blue

above, and the whole set in a large oval patch of rich brownish black, the widest portion of it being above the eye-spot,

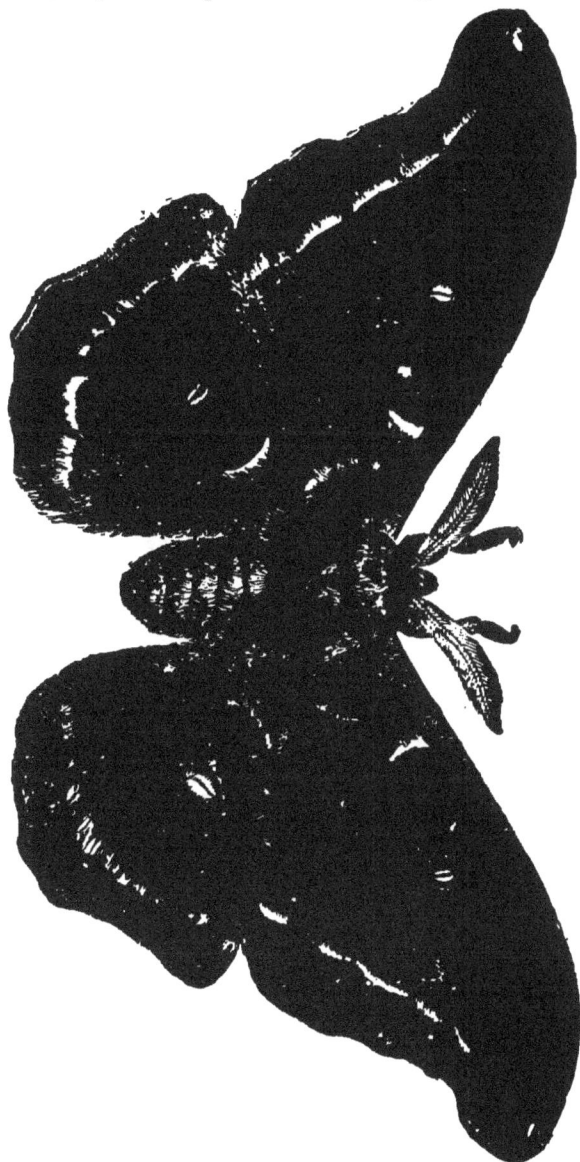

Fig. 186.

where it is sprinkled also with bluish atoms. The front edge
of the fore wings is gray. This lovely creature flies only at

night, and, when on the wing, is of such a size that it is often mistaken for a bat.   Within a few days the female deposits

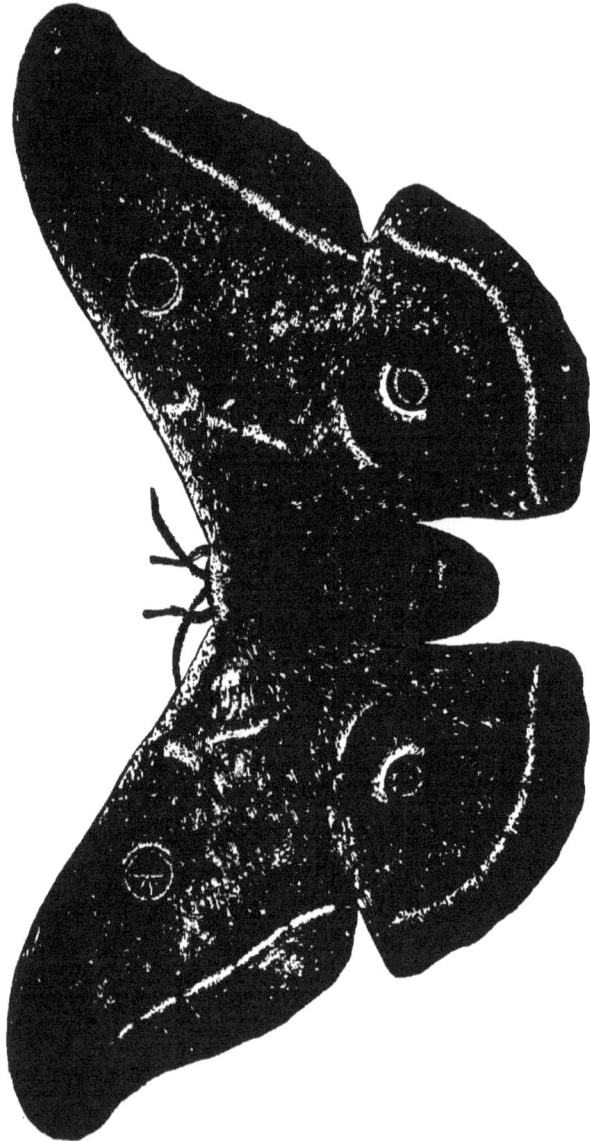

FIG. 186.

her eggs, gluing them singly to the under side of the leaves, usually only one on a leaf, but occasionally two or even three may be found on the same leaf.

The egg is about one-tenth of an inch in diameter, slightly convex above and below, the convex portions whitish, and the nearly cylindrical sides brown. Each female will lay from two to three hundred eggs, which hatch in ten or twelve days.

Remedies.—This insect is subject to the attack of many foes, particularly while in the larval state. A large number fall a prey to insectivorous birds, and they also have insect enemies. An Ichneumon fly, *Ophion macrurum*, the same as that which preys on the Cecropia emperor moth, No. 28 (see Fig. 73), is a special and dangerous foe. This active creature may often be seen in summer on the wing, searching among the leaves of shrubs and trees for her prey. When found, she watches her opportunity, and places quickly upon the skin of her victim a small oval white egg, securely fastened by a small quantity of a glutinous substance attached to it. This is repeated until several eggs are placed, which in a few days hatch, when the tiny worms attach themselves to the skin of the caterpillar and feed on the juices of their victim. The polyphemus caterpillar continues to feed and grow, and usually lives long enough to make its cocoon, when, consumed by the parasites, it dies; in the mean time the Ichneumons, having completed their growth, change to pupæ within the cocoon, and in the following summer, in place of the handsome moth, there issues a crop of Ichneumon flies. The polyphemus caterpillar is also subject to the attacks of another parasite, a Tachina fly. Should the insect ever appear in sufficient numbers to prove troublesome, it can be readily subdued by hand-picking. Besides the plum, the larva feeds on a variety of trees and shrubs, such as oak, hickory, elm, basswood, walnut, maple, butternut, hazel, rose, etc.

No. 89.—The Waved Lagoa.

Lagoa crispata Packard.

The larva of this species is nearly oval, about three-fourths of an inch long, covered above with brownish, evenly-shorn hairs, which are raised to a ridge along the middle of the back, and sloped off on each side like the roof of a house.

It reaches maturity during September, when it makes a tough, oval cocoon, fastened to the side of a twig of the plum-tree on which it has been feeding, and within this changes to a brown chrysalis. The following July the top of the case is opened by the lifting of a flat, circular lid, and from it escapes a pretty moth.

The moth is of a straw-yellow or yellowish-cream color, the fore wings more or less dusky on the outer margin, and covered with fine, flattened, curled hairs, arranged in regular waves, running from near the base to the tip. The wings, when expanded, measure about one and three-quarter inches across. The body and legs are thick and woolly, and at the tip of the abdomen there is a tuft of long, soft hairs, forming a bushy tail. It is common in the South and West, but is not often found in the North; being a comparatively rare insect, it is never likely to give much trouble to the fruit-grower. It is found also feeding on the leaves of the apple and blackberry.

No. 90.—The Streaked Thecla.

Thecla strigosa Harris.

This is a very rare insect, a small butterfly which has never been known to inflict any material damage, but, since its larva has been found feeding on the leaves of the plum-tree, it is deserving of mention.

The caterpillar, when full grown, is half an inch or more in length, of a rich velvety green color, with a tinge of yellow; there is a stripe of a darker shade down the back, with a faint, broken, yellowish line along the middle. The upper part of the body is flattened, the sides abruptly inclined, and striped with faint, oblique, yellowish lines.

When mature, it forms a short, blunt, brown chrysalis, which in ten or twelve days produces the butterfly.

This measures, when its wings are expanded, an inch or more across (see Fig. 187). It is of a plain, dark-brown color above, but beneath the wings are prettily ornamented with wavy white streaks. There is also a row of orange-colored, crescent-shaped spots on the hinder portion of the posterior wings, and a large blue spot near their hind angle. Each of the hind wings has two thread-like tails, one longer than the other.

FIG. 187.

No. 91.—The Plum-tree Catocala.

Catocala ultronia Hubn.

About the middle of June, when jarring the plum-trees for curculios, a very curious-looking, leech-like caterpillar often drops on the sheet spread beneath. It is flattened, with its body thick in the middle and tapering towards each end, and of a grayish-brown color. When full grown, it closely resembles Fig. 188; it is a little more than an inch and a half long,

FIG. 188.

dull grayish brown above, with two or four small reddish tubercles on each segment of the body, all encircled by a slight ring of black at their base. On the upper part of the ninth segment there is a stout, fleshy horn, about one-twelfth of an inch long, pointed, and similar in color to the body, but with

12

an irregular grayish patch on each side. On the twelfth
segment there is a low, fleshy ridge, tinted behind with deep
reddish brown; there is also an oblique stripe on this segment
of the same color, extending forward. Along the sides of the
body, and close to the under surface, there is a thick fringe
of short, fleshy-looking hairs of a delicate pink color. The
under side is also pink, deeper in color along the middle, with
a central row of nearly round black spots, which are largest
from the seventh to the eleventh segment inclusive. The
anterior segments are greenish white, tinted with rosy pink
along the middle.

About the third week in June this larva becomes full
grown, when, fastening together a few leaves with some
silken fibres, it changes within this enclosure to a brown
chrysalis, from which the perfect insect escapes in about three
weeks.

The moth (Fig. 189) has the fore wings of a rich umber

Fig. 189.

color, darkest on the hind margin, with a broad, diffused ash-
colored band along the middle, not extending to the apex,
which is brown. There are also several zigzag lines of brown
and white crossing these wings. The hind wings are deep
red, with a wide black band along the outer margin, and a
narrower band of the same color across the middle. The moth
is on the wing during the greater part of July and August,
during which period the eggs are deposited for the succeeding
brood.

Two other moths have been observed devouring plum leaves, but not in sufficient numbers to attract much attention. The first is *Lithacodes fasciola* Boisd., the larva of which is small, of a uniform green color, and spins a small, oval, brown cocoon between the leaves. The moth is shown in Fig. 190. The other is a tufted caterpillar, the larva of *Parorgyia parallela* G. & R.; it is densely covered with light-brown hairs, and has two black pencils of long hairs projecting in front of the head, and a single tuft of a similar character on the hinder portion of the body.

FIG. 190.

No. 92.—The Leaf-cutting Bee.

Megachile brevis Say.

This is a four-winged fly belonging to the *Hymenoptera*, a species of bee, which curls up the leaves of the plum-tree,

FIG. 191.

and further disfigures it by cutting circular pieces out of other leaves to line the coils and form chambers within them, in which its eggs are deposited, and where the larvæ remain until they reach maturity. The larvæ do not feed on the leaves, but on pollen, or bee-bread, stored up in their cells by the parent insects. This bee is not very abundant, and is never likely to prove very injurious. It is represented in Fig. 191, with examples of the injury it does.

No. 93.—The Plum-tree Aphis.

Aphis prunifolii Fitch.

This aphis resembles in its appearance and habits the apple-tree aphis, No. 57; it is, however, much less common. It infests the under side of the plum leaves, puncturing them and sucking their juices, causing them to become wrinkled and twisted. When first hatched, these insects are of a whitish color tinged with green, but as they increase in size they become of a deeper green, and when mature some of them are black, with pale-green abdomens and dusky wings. The remedies given under the apple-tree aphis (No. 57) are equally applicable to this species.

ATTACKING THE FRUIT.

No. 94.—The Plum Curculio.

Conotrachelus nenuphar (Herbst).

This insect is without doubt the greatest enemy the plum-grower has to contend with, for when allowed to pursue its course unchecked it often destroys the entire crop. The perfect insect is a beetle belonging to a family known under the several names of curculios, weevils, and snout-beetles. It is a small, rough, grayish or blackish beetle, about one-fifth of an inch long (shown, magnified, at *c* in Fig. 192), with a black, shining hump on the middle of each wing-case, and behind this a more or less distinct band of a dull ochre-yellow color, with some whitish marks about the middle; the snout is rather short. The female lays her eggs in the young green fruit

shortly after it is formed, proceeding in the following manner.
Alighting on a plum, she makes with her jaws, which are at
the end of her snout, a small
cut through the skin of the fruit,
then runs the snout obliquely
under the skin to the depth
of about one-sixteenth of an
inch, and moves it backward
and forward until the cavity is
smooth and large enough to receive the egg to be placed in it.
She then turns round, and, dropping an egg into it, again turns
and pushes it with her snout to

FIG. 192.

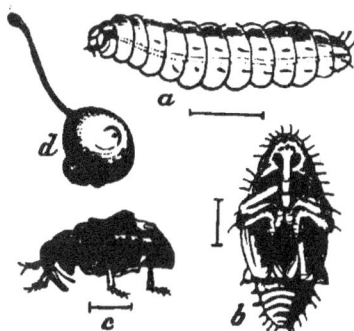

the end of the passage. Subsequently she cuts a crescent-shaped
slit in front of the hole, as shown at *d*, so as to undermine the
egg and leave it in a sort of flap, her object, apparently,
being to wilt the piece around the egg and thus prevent the
growing fruit from crushing it. The whole operation occupies
about five minutes. The stock of eggs at the disposal of a
single female has been variously estimated at from fifty to
one hundred, of which she deposits from five to ten a day,
her activity varying with the temperature.

The egg is of an oblong-oval form, of a pearly-white color,
and large enough to be distinctly seen with the naked eye.
By lifting the flap with the finger-nail or with the point of a
knife it can be readily found. In warm and genial weather
it will hatch in three or four days, but in cold and chilly
weather it will remain a week or even longer before hatching.

The young larva is a tiny, soft, footless grub, with a horny
head. It immediately begins to feed on the green flesh of
the fruit, boring a tortuous channel as it proceeds, until it
reaches the centre, where it feeds around the stone. It attains
its full growth in from three to five weeks, when it is about
two-fifths of an inch long, of a glassy yellowish-white color,
with a light-brown head, a pale line along each side of the

body, a row of minute black bristles below the lines, a second row, less distinct, above, and a few pale hairs towards the hinder extremity. At *a*, Fig. 192, it is shown magnified. The skin of the larva being semi-transparent, the color of the internal organs shows through, imparting to the central portions of the body a reddish hue. The irritation arising from the wound and the gnawing of the grub causes the fruit to become diseased and gummy, and it falls prematurely to the ground, generally before the larva is quite full grown. Within the fallen plum the growth of the larva is completed, when, forsaking the fruit it has destroyed, it enters the ground, burying itself from four to six inches deep, where, turning round and round, it compresses the earth on all sides, until a smooth oval cavity is formed, within which, in a few days, the larva changes to a pupa, shown, enlarged, at *b*, Fig. 192, and in from three to six weeks is transformed to a beetle, which is at first soft and of a reddish color, but soon hardens, and, assuming its natural hue, makes its way through the soil to the surface and escapes.

The insect is single-brooded, the beetle hibernating in secluded spots, under the loose bark of trees and in other suitable places. About the time the plum-trees blossom the curculios are on the alert, and as soon as the fruit is formed the work of destruction begins. Both males and females puncture the fruit to feed on it, but only the females make the peculiar crescent-shaped marks described. They are much more numerous during the early part of the season than later on, and when the weather is warm they are active at night, and deposit eggs then as well as in the daytime. During the middle of the day, and also on warm nights, the beetle readily takes wing; it is less active during the morning and evening. Besides the plum, the peach, nectarine, and apricot also suffer much from its attacks, and it is very injurious to the cherry. In this latter case the infested fruit remains hanging on the tree, and the presence of the enemy is often unnoticed. The beetle also occasionally deposits its eggs in the pear and apple, but in these fruits it seldom matures: either the egg fails to hatch, or the young larva perishes soon after hatching. This insect is native to this country, and has in the past fed on the wild plums, on which it may still be found in considerable numbers.

Remedies.—When the plum curculio is alarmed, it suddenly folds its legs close to its body, turns its snout under its breast, and falls to the ground, where it remains motionless, feigning death. Advantage is taken of this peculiarity to catch and destroy the insect: a sheet is spread under the tree, and the tree and its branches are suddenly jarred, when the beetles fall on the sheet, where they may be gathered up and destroyed. A convenient form of sheet may be made with two or four widths of cotton (depending on the size of the tree), and of the requisite length, stitched only half-way up the middle, to allow the trunk of the tree to pass to the centre, and having each of the sides tacked to a long strip of wood, about an inch square, so that the sheet may be conveniently handled and spread. Small trees may be jarred with the hand ; larger ones should have a branch cut off, leaving a stump several inches long, which may be struck with a mallet, or a hole may be bored in the trunk and a broad-headed iron spike inserted, which is to be struck with a hammer, avoiding as far as practicable any bruising of the bark. As it is important to catch as many of the beetles as possible before any mischief is done, jarring should be begun while the trees are in blossom, and continued daily, morning and evening, if the insects are abundant, for three or four weeks, or until they become very scarce. A form of curculio-catcher, known as Dr. Hull's, is an excellent contrivance where a large orchard has to be cared for. It consists of a wheelbarrow on which is mounted a large inverted umbrella, split in front to receive the trunk of the tree, against which the machine, which is provided with a padded bumper, is driven with force sufficient to jar the curculios down into the umbrella, where they are collected and destroyed. It is very inportant that the fallen plums should be promptly gathered and burnt or scalded, so as to destroy the larva before it has time to escape.

Another remedy, which is less laborious and has been found very effectual, is to syringe or spray the plum-trees as soon as the young fruit has formed with a mixture of Paris green and water, in the proportion of a teaspoonful of the poison to two gallons of water, and repeating the application after a week or ten days. If the weather is very showery, a third spraying may be necessary. This remedy either poisons the curculios or is obnoxious to them, so as to deter them from working on trees so protected. When alternate trees in a plum-orchard where the curculio is common are so treated, the protecting influence of the Paris green is very marked.

Many other remedies have been suggested, but they are all of little value compared with those already given. One of these is to place hogs in plum and peach orchards to devour the fallen fruit; and it is said to have proved in some instances a very successful and inexpensive way of disposing of a large portion of the curculios. Hens with their broods of chickens enclosed within the plum-orchard will devour a large number of the larvæ of the curculio. Hanging bottles of sweetened water on the trees to attract the beetles, scat-

Fig. 193.

tering air-slaked lime through the foliage, and smoking it by burning tar occasionally under the trees, have also been advised. Plum-orchards should not be planted near a wood, as the curculios find shelter there, and are likely to be more numerous than in more open ground; also avoid giving shelter, by removing and burning all rubbish that may accumulate about the trees.

There are many insects which devour the curculio larva as it escapes from the fruit, while some eat into the fruit as it lies upon the ground, seize the culprits, drag them out, and eat them. Foremost among these beneficial insects are two or three species of common ground-beetles belonging to the *Carabidæ;* of these the Pennsylvania ground-beetle, *Harpalus Pensylvanicus* (De Geer), is by far the most common, and may be met with at all

FIG. 194.

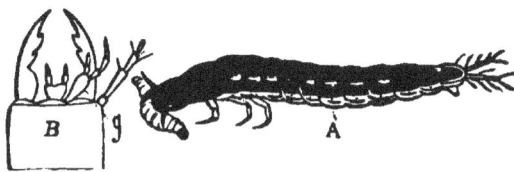

times during the season. Fig. 193 shows it somewhat magnified, and Fig. 194 represents the larva of the same insect, of the natural size, in the act of devouring a curculio larva ; at *b* its formidable jaws are shown, magnified. Fig. 195 shows a larva of one of the larger species of this useful family, magnified.

FIG. 195.

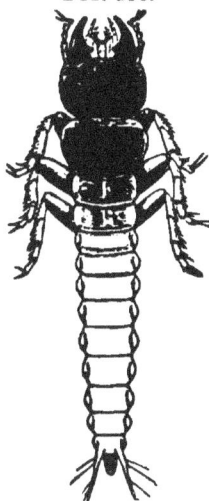

The larva of the soldier-beetle, *Chauliognathus Americanus* (Forst.), is also a useful agent in destroying the curculio. It is shown at *a*, Fig. 196, and a magnified

FIG. 196.

FIG. 197.

view of its head and jaws at *b*. This little friend often works its way into the fruit in search of its prey, sometimes entering it while still on the tree. The perfect beetle (Fig. 197) may be found during the summer on the flowers of the golden-rod, *Solidago*. The larvæ of the lace-wing flies, of the genus *Chrysopa*, one of which is shown in Fig. 132, also devour them ; and ants have been known to destroy the grubs as they leave the fruit to enter the ground. A minute yellow Thrips, scarcely one-twentieth of an inch long, is

FIG. 198.

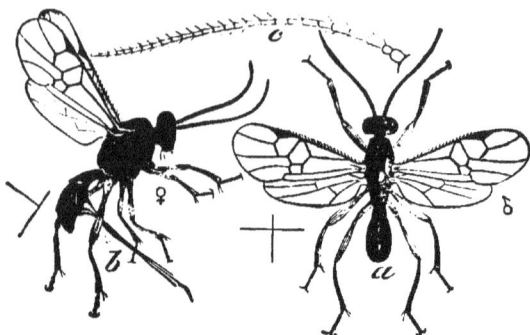

said to seek out and devour large quantities of the eggs of the curculio.

FIG. 199.

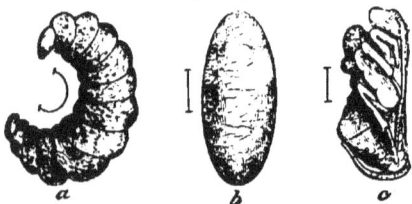

Two species of parasites are known to attack the larva of this pest. One, known as the Sigalphus curculio parasite, *Sigalphus curculionis* Fitch, is a small, black, four-winged fly, represented. in Fig. 198, where *a* shows the male, and *b* the female. With her sharp ovipositor the female punctures the skin of the curculio larva, and deposits an egg underneath, which in due time produces a larva, as shown at *a*, Fig. 199. When the curculio larva is destroyed by the parasite, the latter encloses itself in a small, tough cocoon of yellowish silk, *b*, and then gradually assumes the pupa state, as shown at *c ;* all these figures are magnified. The other species, known as the Porizon curculio

FIG. 200.

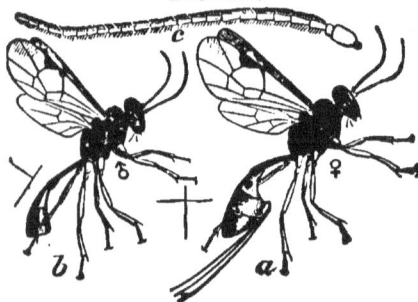

parasite, *Porizon conotracheli* Riley, is also an Ichneumon fly, with similar habits and of about the same size as the species just referred to. In Fig. 200, *a* represents the female, and *b* the male, both magnified. Neither of these parasites has yet appeared in sufficient numbers to act as an efficient check on the increase of the plum curculio.

No. 95.—The Plum-gouger.
Coccotorus scutellaris (Lec.).

While this insect has some points of resemblance to the plum curculio, it is in other respects so different as to be easily distinguished. The beetle, which is shown magnified in Fig. 201, is about five-sixteenths of an inch long, with the thorax and legs of an ochre-yellow color, while the head and wing-cases are brown, with a leaden-gray tint, the latter with whitish and black spots scattered irregularly over their surface. The wing-cases are without humps; the snout is somewhat longer than the thorax, and projects forward or downward, but cannot be folded under the breast as in the case of the plum curculio. It appears in spring about the same time as the plum curculio, but, instead of making a crescent-shaped slit in the plum, it bores a round hole like the puncture of a pin.

Fɪɢ. 201.

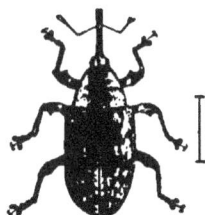

The eggs are deposited in the following manner. With the minute but powerful jaws at the tip of the snout of the female, a hole is made about four-fifths as deep as the snout is long, which is enlarged at the end and gouged out somewhat in the form of a gourd. The egg is placed in the excavation, and pushed down with the snout until it reaches the receptacle prepared for it. After being deposited, it swells from absorption of the surrounding moisture, and within a few days the young larva escapes.

On escaping from the egg, it makes an almost straight course for the kernel of the plum, through the soft shell of which it makes its way, and feeds upon the contents until full grown. When nearly mature, the larva, by a wise instinct, prepares a way for the escape of the future beetle by cutting a round hole through the now hard stone. The larva is of a milk-white color, with a large, horny, yellowish-white head, and jaws tipped with brown. It enters the chrysalis state within the plum-stone, and, when mature, the beetle passes through the hole bored by the larva, makes its way through the flesh, and escapes.

While the normal habit of the plum curculio is to feed on the flesh outside the plum-stone, which latter it only occasionally penetrates, the plum-gouger lives and matures within. Both sexes of the plum-gouger bore cylindrical holes in the fruit for food ; and where the insect abounds, the growing fruit will be found covered with these punctures, from which more or less gum exudes, and the fruit becomes knotty and worthless, but does not readily drop, as do those which have been injured by the plum curculio. The insect is single-brooded, and requires a longer time to mature than the plum curculio ; eggs deposited in June do not produce beetles until the end of August or early in September. It appears to be unknown in the Eastern States, but is very generally distributed throughout the valley of the Mississippi. It is much less common, and does far less injury, than the plum curculio, although occasionally it is found in almost equal abundance. It is said to pass the winter in the beetle state.

Remedies.—This beetle may be collected by jarring the trees in the manner described for the plum curculio, although it does not drop quite so readily ; it also takes wing quickly, and hence is not so easily secured.

No. 96.—The Saddled Leaf-hopper.

Bythoscopus clitellarius Say.

This insect is occasionally injurious to the plum, by puncturing the stems of the fruit and sucking the fluids which should go to nourish and mature it. It is a small leaf-hopper (shown in Fig. 202), about one-fifth of an inch long, of a dark-brown or black color, with a sulphur-yellow spot like a saddle upon the middle of its back, and in front of this a band of pale yellow,—the head and under side being of the same color. It is unlikely that this insect will ever occur in sufficient numbers to cause much injury.

FIG. 202.

SUPPLEMENTARY LIST OF INJURIOUS INSECTS WHICH AFFECT THE PLUM.

ATTACKING THE ROOTS.

The peach-tree borer, No. 97, sometimes invades the plum-tree, and burrows about the collar and into the larger roots adjacent without causing an exudation of gum, as in the peach. Young trees are most liable to injury.

ATTACKING THE TRUNK.

The flat-headed apple-tree borer, No. 3, frequently attacks the plum and materially injures the tree.

ATTACKING THE LIMBS AND BRANCHES.

The parallel Elaphidion, No. 12; the pear-blight beetle, No. 68; the New York weevil, No. 100; and the tree-cricket, No. 178.

ATTACKING THE LEAVES.

The apple-tree tent-caterpillar, No. 20; the forest tent-caterpillar, No. 21; the white-marked tussock-moth, No. 22; the canker-worms, Nos. 25 and 26; the fall web-worm, No. 27; the Cecropia emperor moth, No. 28; the unicorn prominent, No. 29; the blind-eyed sphinx, No. 31; the oblique-banded leaf-roller, No. 35; the leaf-crumpler, No. 37; the eye-spotted bud-moth, No. 38; the tarnished plant-bug, No.

71; the pear-tree slug, No. 75; the May-beetle, No. 113; the Ursula butterfly, No. 116; the basket-worm, or bag-worm, No. 120; the pyramidal grape-vine caterpillar, No. 147; the grape-vine flea-beetle, No. 150; the rose-beetle, No. 151; and the currant Amphidasys, No. 211, all devour the leaves, while the pear-tree blister-beetle, No. 73, eats both leaves and blossoms.

ATTACKING THE FRUIT.

The codling moth, No. 58, occasionally injures the fruit; so, also, do bees and wasps, when it is fully ripe.

ATTACKING THE TRUNK.

No. 97.—The Peach-tree Borer.

Ægeria exitiosa Say.

This notorious pest, so destructive to peach-orchards, is very widely disseminated. The parent insect belongs to a family of moths known as Ægerians, which, having transparent wings and slender bodies, strongly resemble certain wasps and hornets, and, as they fly in the daytime only, and are then very active on the wing, the resemblance becomes still more striking. The moth appears in the Northern States and Canada from about the middle of July to the end of August; in the South it appears much earlier,—in some localities as early as the latter part of May. The sexes differ very much in appearance. In Fig. 203, *a* represents the female, and *b* the male. The female is much the larger, and has a broad, heavy abdomen. The body is of a glossy steel-blue color with a purplish reflection, and a broad band of orange-yellow

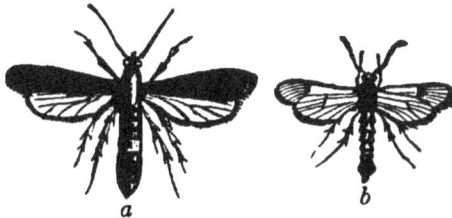

FIG. 208.

across the abdomen. The fore wings are opaque, and similar in color to the body, their tips and fringes having a purplish tint both above and beneath. The hind wings are transparent and broadly margined with steel-blue; when the wings are expanded, the moth measures about an inch and a half across. The male is smaller, its wings seldom measuring more than an inch; its body, which is also of steel-blue color, with golden-yellow markings and a glossy, satin-like lustre, is much more slender than that of the female. The antennæ are black and densely fringed on the inner side with numerous fine, short hairs, the latter a feature absent in the female. The head and thorax are marked with yellow, and the abdomen has two slender yellow bands above, and a white line on each side of the tuft of hairs at its tip. The wings are transparent, the veins, margins, and fringe steel-blue, and a steel-blue band extends nearly across beyond the middle. The feet and legs are marked with yellow and white.

The female deposits her eggs on the bark of the tree at the surface of the ground. They are about one-fiftieth of an inch long, with a sculptured surface, oval in form, slightly flattened, and of a dull-yellowish color. They are deposited singly, are fastened to the surface of the bark by a gummy secretion, and sometimes have a few of the dark-blue scales from the tip of the abdomen of the female attached to them.

As soon as the larva is hatched, it works downwards in the bark of the root, forming a small winding channel, which soon becomes filled with gum. As it increases in size, it devours the bark and sap-wood, and causes a copious exudation of gum, which eventually forms a thick mass around the base of the tree, intermingled with the castings of the worm. When full grown (see Fig. 204), the larva measures over half an inch in length, and nearly a quar-

FIG. 204.

ter of an inch in diameter. It is a naked, soft, cylindrical grub, of a pale whitish-yellow color, with a reddish, horny-looking head and black jaws; the upper part of the next segment is similar in appearance to the head, but of a paler shade. The under surface resembles the upper in color; the three anterior pairs of claw-like feet are tipped with brown; the five hinder pairs of thick, fleshy prolegs are yellow, each of the latter margined with a fringe of very minute reddish-brown hooks. There are a few scattered hairs over the surface of the body, each arising from a pale-reddish, wart-like dot. The larvæ may be found of different sizes all through the fall and winter months, some quite young associated with others nearly full grown. During the winter the larger ones rest, with their heads upwards, in smooth, longitudinal grooves which they have excavated, the back part being covered with castings mingled with gum and silken threads, forming a kind of cell, the cavity of which is considerably larger than the worm inhabiting it; the smaller ones usually lie in the gum, or between it and the wood of the trunk or root. In badly-infested trees the whole of the bark at the base or collar is sometimes consumed for an inch or two below the surface. (Nor does the insect always confine itself to the base of the tree; occasionally it attacks the trunk farther up, and sometimes the forks of the limbs; but the exuding gum invariably points out the spot where the foe is at work.)

When about to become a pupa, the larva crawls upwards to the surface of the ground, and constructs a pod-like case, of a leathery structure, made from its castings mixed with gum and threads of silk. It is about three-quarters of an inch long, of a brown color, oval in form, with its ends rounded ; its inner surface is smooth, and it is fastened against the side of the root, often sunk in a groove gnawed for that purpose, with its upper end protruding slightly above the surface of the ground. If the earth has recently been disturbed about the surface of the tree, so as to make it lie loose, the larva will often form its cocoon an inch or more below the surface. The enclosed pupa is at first white, but soon becomes of a pale tawny-yellow color, with a darker ring at each of the sutures of the body ; the pupa state lasts some three weeks or more.

This is an American insect, unknown on the peach-trees of other countries. Its operations are not confined to the peach ; it works also on the plum, although in this instance no gum exudes from the tree, and it is quite probable that before the introduction of the peach into this country the larva lived in the roots of the wild plum, which it has now almost entirely forsaken.

18

Remedies.—Several remedies have been proposed to meet this evil. Where the larvæ are present, they are readily detected in consequence of the exudation of gum; hence early in spring the trees should be carefully examined, a little of the earth removed from about the base, and, if masses of gum are found, the larvæ searched for and destroyed. Hot water is said to be very effectual in killing them; it should be used very hot, and after the earth has been removed, so as to insure its reaching the culprits before it cools. Among the preventive measures, much has been written in favor of mounding the trees, banking the earth up around the trunk to the height of a foot or more, and pressing it firmly about the tree. Some allow the mounds to remain permanently, but the better plan seems to be to mound up late in the spring or towards midsummer, and level off the ground again in September, after egg-laying has ceased and the moths have disappeared. This treatment is said to make the bark very tender and liable to injury during the winter, and it is recommended by some to defer its application until the fourth year, by which time the bark will have become sufficiently thickened and hardy to endure the treatment without injury. Placing around the roots a bed of cinders, ashes, or lime, plastering the base of the trunk with mortar or clay and covering it with stout paper, coating the tree with an application of soap or tobacco-water, have all had their advocates; but the weight of testimony is in favor of the removal of the larvæ with the knife late in the autumn or early in the spring, and subsequently mounding the trees in the manner already described.

Another remedy proposed is to cover the trunk with straw in the following manner. Scrape the earth away from the collar, place a handful of straight straw erect around the trunk, fastening it with twine, then return the soil, which will keep the ends of the straw in their place. The straw should entirely cover the bark, and the twine be loosened as the trunk increases in size. Trees so protected are said to have remained uninjured while all around them have suffered from the borer.

No. 98.—The Elm-bark Beetle.
Phlœotribus liminaris (Harris).

This insect is very common on elm-trees; it also occasionally attacks the peach-tree, especially when from any cause it has become diseased. In August or September there appear small perforations like pin-holes in the bark, from which issue minute cylindrical beetles about one-tenth of an inch long, of a dark-brown color, with the wing-cases deeply impressed with punctated furrows, and covered with short hairs; the thorax is also punctated. This species has never occurred on the peach in sufficient numbers to attract general attention, or to require the adoption of any special remedies.

ATTACKING THE BRANCHES.

No. 99.—The Peach-tree Bark-louse.
Lecanium persicœ (Fabr.).

This is an insect very similar in appearance and habits to the pear-tree bark-louse, No. 69. It is found attached to the smooth bark of the peach twigs, frequently beside a bud or at the base of a twig, appearing as a black hemispherical shell about the size and shape of a split pea; its surface is uneven, shining, commonly showing a pale margin, and a stripe upon the middle. It feeds upon the sap, piercing the bark with its proboscis, and imbibing the juices. When mature, the removal of the scale discloses a multitude of eggs, which in due time hatch, and the young larvæ scatter over the twigs, and, fastening themselves to the bark, become permanently located, and live the full term of their lives without changing their position.

No. 100.—The New York Weevil.

Ithycerus noveboracensis (Forster).

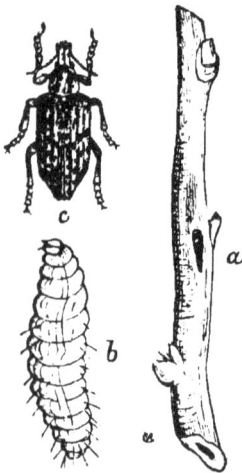

This is a snout-beetle or curculio, the largest species we have in this country. It appears in May or June, and injures fruit-trees by eating the buds and gnawing into the twigs at their base, often causing them to break and fall; it also gnaws off the tender bark early in the season before the buds have expanded, and later eats the leaves off just at their base, and devours the tender shoots. It is from four to six tenths of an inch in length (see *c*, Fig. 205), of an ash-gray color marked with black; on each of its wing-cases there are four whitish lines interrupted by black dots, and three smaller ones on the thorax. The scutel, which is at the point of junction of the wing-cases with the thorax, is yellowish. The beetle is said to be more active at night than in the day, and seems to show a preference for the tender, succulent shoots of the apple, although it makes quite free with those of the peach, pear, plum, and cherry. Sometimes it occurs in swarms in nurseries, when it seriously injures the young trees. In the East it is seldom present in sufficient numbers to prove injurious, but it is very common in the valley of the Mississippi. The larva is found in the twigs and tender branches of the bur-oak, and probably also in those of the pig-nut hickory.

Fig. 205.

When the female is about to deposit an egg, she makes a longitudinal excavation with her jaws, as shown at *a* in Fig. 205, eating upwards under the bark, and afterwards turns round and places an egg in the opening.

The larva (*b* in the figure) is a soft, footless grub, of a pale-yellow color, with a tawny head; it is not known whether it undergoes its transformations within the twig, or enters the ground to pass the pupa state.

Remedies.—There seems to be none other than to catch and kill this mischief-maker. In common with almost all other curculios, this beetle has the habit of falling to the ground when alarmed, and hence may be captured by jarring the trees in the manner directed for the plum curculio, No. 94.

ATTACKING THE LEAVES.

No. 101.—The Peach-tree Leaf-roller.
Ptycholoma persicana (Fitch).

Early in spring, when the young leaves are expanding, a small worm sometimes attacks them, and, drawing them together with fine silken threads, secretes itself within, and feeds upon them. This larva is rather slender, of a pale-green color, with a pale, dull-yellowish head, and a whitish streak along each side of its back. When full grown, it changes to a chrysalis within its nest, where it remains about two weeks, and then escapes as a moth.

The fore wings of the moth are of a reddish-yellow color, varied with black; at the base they are paler; there is a large, white, triangular spot on the middle of the outer margin, and a transverse streak of the same hue within the hind margin. This latter is divided by the veins crossing it into about four spots, and is bordered on its anterior side by a curved black band. When its wings are spread, this moth measures nearly three-quarters of an inch across. It has never yet been reported as very destructive anywhere, and is scarcely likely to require the application of any special remedy.

No. 102.—The Blue-spangled Peach-tree Caterpillar.
Callimorpha Lecontei Boisd., *var. fulvicosta* Clem.

Very early in spring there may sometimes be found sheltered under the loose bark of peach-trees, and sometimes also

on apple-trees, small black caterpillars covered with short
stiff hairs and studded with minute blue spots. As soon as
the leaves begin to expand, these larvæ issue from their hiding-
places and feed upon them. They grow rapidly, and soon
attain their full size, when they are nearly an inch long, and
appear as shown at *a*, Fig. 206; *c* shows an enlarged side

Fig. 206.

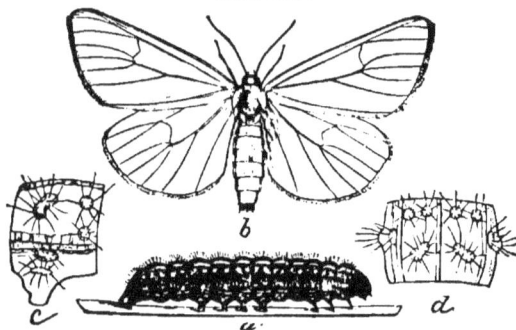

view of one of the segments of the body, and *d* a back view
of the same. The full-grown caterpillar is of a velvety
black color above, and pale bluish, speckled with black, below.
There is a deep orange line along the back, and a more distinct
wavy and broken line along each side. The warts from which
the bristly hairs issue are of a steel-blue color, with a polished
surface, which reflects the light so as to make them appear
quite brilliant.

The larva selects some sheltered spot and there spins a slight
cocoon of white silk, within which it changes to a chrysalis
of a purplish-brown color, finely punctated, and terminating in
a flattened plate tipped with yellowish-brown, curled bristles.

The moth issues during the early part of June in the
Northern and Middle States; it is of a milk-white or cream
color, with the head, collar, and base and tip of the abdomen
orange-yellow. On the under side the anterior margins of
the wings, the legs, and the body partake of the same hue.
When spread, the wings measure about one and three-quarter
inches across.

Remedies.—When these larvæ are numerous they sometimes do considerable damage to the young foliage of the peach-tree. They may be subdued by hand-picking, or by shaking them from the trees and crushing them under foot, or by syringing the leaves of the trees with Paris-green and water in the proportion of a teaspoonful to two gallons of water.

No. 103.—The Peach-tree Aphis.
Myzus persicæ Sulzer.

This aphis begins to work upon the young leaves of the peach-trees almost as soon as they burst from the bud, and continues throughout the greater part of the season, unless swept off, as sometimes happens with surprising rapidity, by insect enemies. These lice live together in crowds under the leaves, and suck their juices, causing them to become thickened and curled, forming hollows with corresponding reddish swellings above; frequently the curled leaves fall prematurely to the ground. The perfect winged females are about one-eighth of an inch long, black, with the under side of the abdomen dull green, the wingless females rusty red, with the antennæ, legs, and honey-tubes greenish. The winged males are bright yellow, streaked with brown, with black honey-tubes.

The insects which prey on the apple-tree aphis, No. 57, feed on this species also, and the remedies recommended for that insect are equally applicable to this one.

SUPPLEMENTARY LIST OF INJURIOUS INSECTS WHICH AFFECT THE PEACH.

ATTACKING THE TRUNK.

The flat-headed apple-tree borer, No. 3, and the divaricated Buprestis, No. 104, both injure the trunk of the peach-tree.

ATTACKING THE BRANCHES.

The buffalo tree-hopper, No. 18; the red-shouldered Sin-oxylon, No. 130; the tree-cricket, No. 178; and the strawberry root-borer, No. 190, all attack the branches. The stalk-borer, No. 201, sometimes bores into the buds and young branches.

ATTACKING THE LEAVES.

The oblique-banded leaf-roller, No. 35; the leaf-crumpler, No. 37; the many-dotted apple-worm, No. 43; the saddled leaf-hopper, No. 96; the basket-worm, or bag-worm, No. 120; the rose-beetle, No. 151; and the smeared dagger, No. 194, devour the leaves.

ATTACKING THE FRUIT.

The codling moth, No. 58; the ash-gray pinion, No. 64; the Indian Cetonia, No. 81; and the plum curculio, No. 94, all affect the fruit, the last-named insect being especially injurious.

INSECTS INJURIOUS TO THE APRICOT AND THE NECTARINE.

The nectarine and apricot, being closely related to the peach, are liable to be injured by the same insects; besides those enumerated as affecting the peach, the apricot occasionally suffers in its branches from the attacks of the pear-blight beetle, No. 68.

INSECTS INJURIOUS TO THE CHERRY.

ATTACKING THE TRUNK.

No. 104.—The Divaricated Buprestis.

Dicerca divaricata (Say).

This is a beetle belonging to the family Buprestidæ, most of the members of which are readily distinguished by their coppery or bronzed appearance. This species (see Fig. 207) is from seven to nine tenths of an inch in length, copper-colored, and sometimes brassy, and thickly covered with little indentations. The thorax is furrowed in the middle, and the wing-covers are marked with numerous irregular impressed lines and small, elevated, blackish spots. The wing-cases taper much behind, and their long and narrow tips are blunt-pointed, and spread apart a little,

Fig. 207.

the latter peculiarity having given to the insect its specific name, *divaricata*. The beetles may be found sunning themselves upon the limbs of cherry and peach trees during June, July, and August; they are active creatures, running briskly up and down the trunks of the trees in the sunshine.

The female deposits her eggs on the cultivated and wild cherry-trees, probably in crevices in the bark, and also on the peach, and, when hatched, the young larva bores through the bark and lives in and destroys the sap-wood underneath. It is a flattened larva, with its anterior segments very much enlarged, and closely resembles that of the flat-headed apple-tree borer, No. 3, Fig. 4, but is larger. This insect is seldom very troublesome; should it require attention, the remedies recommended for No. 3 will be equally applicable to this species.

201

No. 105.—The Spotted Horn-beetle.

Dynastes tityus (Linn.).

This is an enormous beetle, some two inches in length, exclusive of its horns. It is of a pale-olive color, with the wing-covers spotted and dotted with black. In the males the middle of the thorax is extended forward in the form of a long black horn, which is hairy along its under side, and

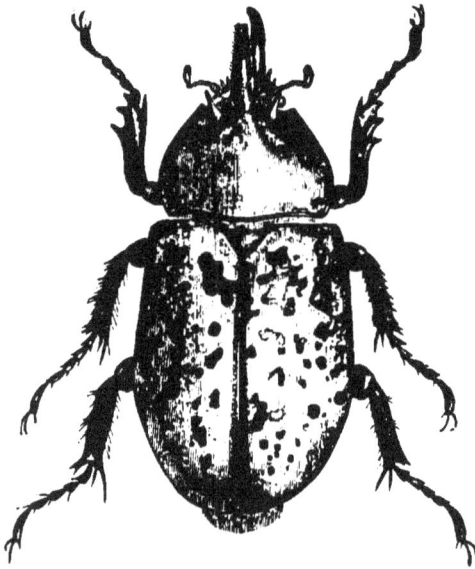

Fig. 208.

usually notched at its tip, as if formed to receive the sharp point of another similar horn, which curves upwards from the crown of the head. There are two other horns between these, short and sharp-pointed. The female is smaller than the male, and unarmed, except with a small tubercle on the head. Fig. 208 represents the male.

The beetle occasionally varies in color: specimens have been found with chestnut-brown wing-covers, others with the thorax black; and in one instance a male was taken with one of the wing-covers black, while the other was of the normal character.

The larva of this insect bores in old, decaying cherry-trees. It somewhat resembles that of the rough Osmoderma, No. 8, but is much larger. The beetle is frequently met with in the South, and is sometimes found as far north as Pennsylvania, but the damage it inflicts is very slight.

ATTACKING THE BRANCHES.

No. 106.—The Dog-day Cicada.

Cicada tibicen Linn.

In appearance this insect very much resembles the seventeen-year locust, No. 15, but differs from it by occurring in more or less abundance every year during the months of August and September, when it sometimes wounds the small limbs of the cherry and deposits its eggs therein. The body is black on the upper side, the head and thorax being spotted and marked with olive-green. The wings are large, transparent, and strongly veined, the principal veins having a greenish tint. The under side of the body is coated with a whitish powder, legs greenish. This cicada, which is shown in Fig. 209, is very generally distributed throughout the Northern United States and the province of Ontario, and the shrill notes of the males may be heard almost everywhere during warm days in August, from ten o'clock in the morning until two in the afternoon. The males only are musical, and their drums are situated in cavities in the sides of the anterior segments of their robust bodies.

FIG. 209.

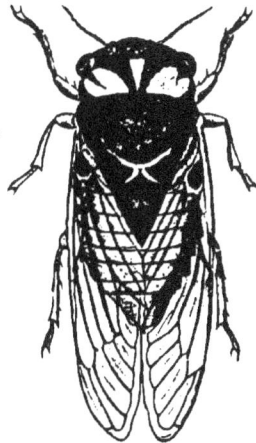

The larva is unknown, but doubtless closely resembles that of the seventeen-year locust; the pupa also is very similar, and has been found beneath cherry, maple, and elm trees. The ravages of this insect have never been sufficiently important to attract much attention.

No. 107.—The Cherry-tree Bark-louse.

Lecanium cerasifex Fitch.

This is a bark-louse very much resembling that of the pear-tree, *Lecanium pyri*, No. 69. It may be found in spring adhering to the under side of the limbs of cherry-trees and sucking their juices. The shell is hemispherical in form, black, more or less mottled with pale dull-yellow dots. On lifting this shell, a mass of minute eggs is found, which shortly hatch, whereupon the insects spread over the bark of the succulent twigs, and, piercing it, subsist upon the juices, passing through the various stages of their growth before the winter approaches. The remedies recommended for *L. pyri* will be equally applicable in this case.

No. 108.—The Cherry-tree Scale-insect.

Aspidiotus cerasi Fitch.

On examining the limbs of the choke-cherry in winter, there will sometimes be found on the bark a small, roundish scale, like a tiny blister, which, when raised, discloses a cluster of very minute dull-reddish eggs, the product of the cherry scale-insect, which is believed to be identical with the scurfy bark-louse, No. 17, and to which the same remedies may be applied.

ATTACKING THE LEAVES.

No. 109.—The Violaceous Flea-beetle.

Crepidodera Helxines (Linn.).

From about the middle of May until August there may often be found on the leaves of cherry-trees small flea-beetles, about one-tenth of an inch long, and of a brilliant coppery, violet, or greenish-black color, with the antennæ of a pale yellow, the under side black, and the legs, except the hinder thighs, dull pale yellow. Though small, this is a very active insect. It gnaws round pieces out of the under side of the leaf, leaving the upper skin unbroken, and sometimes eats entirely through, making numerous small holes in the young leaves at the ends of the limbs. It has not yet proved sufficiently troublesome to require any special remedy.

No. 110.—The Promethea Emperor-moth.

Callosamia Promethea (Drury).

During the winter there may frequently be seen on cherry-trees, particularly the wild species, a twisted leaf hanging here and there after all the others have fallen. A closer examination shows each of these to contain a long, oval, silken cocoon (see Fig. 210), the stem of the leaf being secured to the twig on which it grew with silken threads. The silk is wound round the twig for about half an inch on each side, then carried down around the leaf-stalk to the cocoon, the whole being so firmly fastened that the leaf with the cocoon cannot be detached without much force. This is the cocoon of the Promethea emperor-moth. Besides the cherry, it is found on the sassafras, lilac, button-bush, and occasionally on other trees and shrubs.

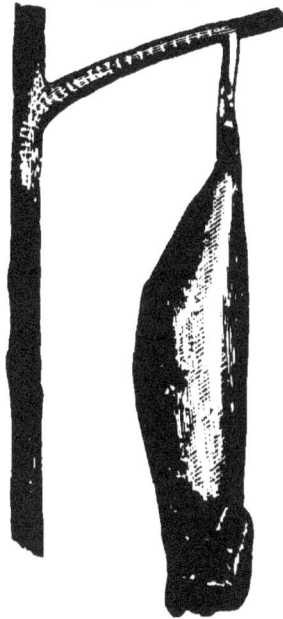

The moth escapes late in June or early in July. It is a handsome insect, and measures, when its wings are expanded, from three and a half to nearly four and a half inches across. The sexes differ very much in appearance: the wings of the male (Fig. 211) are brownish black, those of the female (Fig. 212) light reddish brown. In both, the wings are crossed by a wavy whitish line near the middle, and a clay-colored border along the hind edges. Both also have an eye-like black spot, with a pale-bluish crescent within, near the tip of the fore wings. Near the middle of each of the wings of the female there is an angular reddish-white spot, edged with black; the same is visible on the under side of the wings of the male, but is seldom seen on the upper side.

FIG. 210.

The female lays her eggs in small clusters of five or six or more together; they are of a creamy-white color, about one-

FIG. 211.

sixteenth of an inch in diameter, with an ochreous-yellow spot on the upper side. They hatch towards the end of July.

FIG. 212.

The newly-hatched larva is about one-third of an inch long, pale green, with yellow bands and faint rows of black tubercles. After it has passed the second moult it appears as seen at *a*. From the end of August until late in September it may be found full grown, when it measures two inches or more in length and about half an inch in diameter, and presents the appearance shown at *b* in Fig. 213. It is of a bluish-green or sometimes of a greenish-yellow color, with the head, feet, and hinder segments yellow. There are about eight small warts or short horns of a deep-blue color on each segment, except the two uppermost on the top of the third and fourth segments, which are of a rich coral-red color, and a long one on the top of the twelfth ring, which is yellow.

The caterpillar is found feeding on the cherry, ash, sassafras, poplar, azalea, cephalanthus, or button-bush, and other shrubs and trees. Although the ash is a very common food-plant for the larva, it is rarely, if ever, that a cocoon is found upon it; the leaf-stalks being so very long, it is probably too laborious a task for the caterpillars to fasten them to the twigs, and hence they wander off in search of leaves with shorter stalks and of a thicker, more leathery structure, such as the cherry or the lilac, which form a substantial covering for the cocoon.

The cocoons are often perforated by birds during the winter and their contents devoured. The insect is also subject to the attacks of a small four-winged parasite, a species of Ichneumon.

No. 111.—The Purblind Sphinx.

Smerinthus myops (Sm. & Abb.).

There are sometimes found on cherry-trees, devouring the leaves, in the month of August, large, cylindrical, green larvæ, about two inches long, with a curved horn at the end of the body. The head is bluish green, with a bright-yellow line on the sides; the body is green, with a row of reddish-brown spots on each side of the back, and another similar row lower down near the breathing-pores. Along each side there are six oblique bright-yellow bands, and two short yellow lines on the anterior segments. The horn is green, tinted with yellow at the sides. This is the larva of the purblind sphinx.

When full grown, it buries itself under the ground, where it changes to a dark-brown chrysalis, and in this condition remains until the following June or July, when the perfect insect escapes.

The moth is a very handsome one (see Fig. 214), and measures, when its wings are expanded, about two and a half inches

FIG. 214.

across. The head and thorax are chocolate-brown with a purplish tinge, the thorax having a tawny yellow stripe down the middle; the abdomen is brown, with dull-yellowish spots. The fore wings are chocolate-brown, with black bands and patches, and are angulated and excavated on the hind margin. The hind wings are dull yellow, with the outer half chocolate-brown, and have an eye-like spot towards the inner margin, black, with a large pale-blue centre.

The insect is a rare one, and not likely ever to occur in sufficient numbers to do much injury.

No. 112.—The Io Emperor-moth.

Hyperchiria Io (Linn.).

This very beautiful insect appears in June and July. It remains inactive during the day, but flies about after dusk. The sexes differ in both size and color, the male (Fig. 215)

FIG. 215.

being the smaller. It is of a deep-yellow color, with purplish-brown markings; on the fore wings are two oblique wavy lines near the outer margin, a zigzag line near the base, and other blackish dots and markings. The hind wings are of a deeper ochre-yellow, and are shaded with purple next the body; within the hind margin is a curved purplish band, and inside this a smaller one of a dark-purplish shade, while about the middle of the wing there is a large, round, blue spot with a whitish centre and enclosed in a broad ring of brownish black. The antennæ of the male are beautifully feathered, and the wings measure, when expanded, about two and a half inches across. The female (Fig. 216) measures from three to three and a half inches. The antennæ are but very slightly feathered; the fore wings are purplish brown mingled with gray, the wavy lines crossing the wings being also gray. There is a brown spot about the middle, margined

14

by an irregular gray line, and towards the base the wings are densely clothed with a wool-like covering. The hind wings are very similar to those of the male; the thorax and legs are purplish brown, the abdomen ochre-yellow, with a purplish-red edging on each ring.

Shortly after pairing, the female deposits her eggs in clusters, sometimes as many as twenty or thirty in one group.

Fig. 216.

They are top-shaped, compressed on both sides, and flattened above, about one-sixteenth of an inch long, and one-twentieth of an inch in the longest diameter, creamy white in color, with a yellowish spot above, which gradually becomes darker as it approaches maturity, until it is almost black, when the yellow larva within begins to show through the translucent sides.

The young larvæ are darker in color than the more matured specimens; they keep together in little swarms, and when moving from one place to another follow each other in regular processionary order, a single caterpillar taking the lead, closely followed sometimes by one or two in single file, then by two, three, four, or more, in regular ranks. When about half grown, they lose this habit, and, separating, each one shifts for itself. The larva attains maturity during August, when it measures two and a half inches or more in length and is of a corresponding thickness. (See Fig. 217.) It is of a delicate pale-green color, paler, approaching whitish, along the back, with a broad dusky - white stripe on each side, margined with reddish lilac; breathing-pores yellow, ringed with brown. The body is covered with clusters of green branching spines tipped with black, arising from small warts, of which there are a number on each segment. These spines are very sharp, and when the insect is carelessly handled they sting severely, producing on the more tender portions of

Fig. 217.

the skin an irritation, accompanied by redness and raised white blotches, very similar to that of the stinging nettle. Fig. 218 shows some of these branching spines magnified, *b* being stouter and more acute than the others.

Fig. 218.

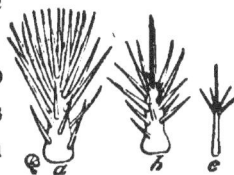

When full grown, the larva descends to the ground, and, drawing together portions of dead leaves or other rubbish to form an outer covering, constructs within this a slight cocoon of tough, gummy, brown silk, in which the change to a chrysalis takes place. The chrysalis is rather short and thick, of a pale-brown color, with a few reddish bristles on the abdominal joints, and a tuft of the same at the end.

While common on the cherry, this caterpillar does not confine itself to one kind of food, but is also found feeding on the apple, thorn, willow, elm, dogwood, balsam poplar, sassafras, locust, oak, currant, clover, cotton, and other plants, shrubs, and trees. It is much more plentiful in some seasons than in others, but, in consequence of its using so many different sorts of food, it is seldom noticed as very injurious to any particular kind of tree, shrub, or plant. Should it prove troublesome, it may easily be subdued by hand-picking, the operator using a pair of gloves while engaged in the work. The larva is attacked by parasites, particularly by a small, undetermined, four-winged fly. The long-tailed Ophion, *Ophion macrurum*, referred to under No. 28 (see Fig. 73), also preys upon it.

No. 113.—The May-beetle.
Lachnosterna fusca (Fröhl.).

Every one must be familiar with the May-beetle,—or May-bug, as it is commonly called,—a buzzing beetle, with a slow but wild and erratic flight, which comes thumping against the windows of lighted rooms in the evenings in May and early in June, and, where the windows are open, dashes in without a moment's consideration, bumping against walls, ceiling, and articles of furniture, occasionally dropping to the floor, then suddenly rising again. It sometimes lands uninvited on one's face or neck, or, worse still, on one's head, where its sharp claws become entangled in the hair in a most unpleasant manner. It is a thick-bodied, chestnut-brown or black beetle (see Fig. 219, 3 and 4), from eight to nine tenths of an inch in length. Its head and thorax are punctated with small indentations; the wing-covers, though glossy and shining, are roughened with shallow, indented points, and upon each there are two or three slightly elevated lines running lengthwise. Its legs are tawny yellowish, and the breast is covered with pale-yellowish hairs; the under surface is paler than the upper. During the day the beetles remain in repose, but are active at night, when they congregate upon cherry, plum, and other trees, devouring the leaves,—occasionally, when very numerous, entirely stripping the trees of foliage. Their strong jaws are well adapted for cutting their food, and their notched or double claws support them securely on the foliage.

The female is said to deposit her eggs between the roots of grass, enclosed in a ball of earth; they are white, translucent, and spherical, and about one-twelfth of an inch in diameter. When hatched, the small white grubs begin at once to feed upon the rootlets of plants; they are several years in reaching maturity, and hence larvæ of different sizes are usually found

FIG. 219.

in the ground at the same time. When full grown, they are almost as thick as a man's little finger; they are soft and white, have a horny head of a brownish color, and six legs; the hinder part of the body is usually curved under, as shown at 2, Fig. 219. This larva is generally known as "the white grub," and is very injurious to strawberries, devouring the roots and destroying the plants; it feeds also upon the roots of grass and other plants, and when very numerous it so injures pasture-lands and lawns that large portions of the turf can be lifted with the hand and rolled over like a piece of carpet, so completely are the roots devoured. When cold weather approaches, the grub buries itself in the ground deep enough to be beyond the reach of frost, and there remains until the following spring.

FIG. 220.

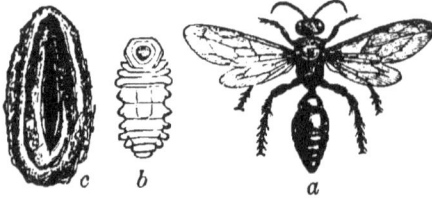

c   b   a

FIG. 221.

When ready for its next change, the larva forms a cavity in the ground, by turning itself round and round and pressing the earth until it moulds a cell of suitable form and size, which it lines with a glutinous secretion, so that the cell may better retain its form, and within this it changes at first to a pupa (shown at 1 Fig. 219), and finally produces the perfect beetle.

Remedies.—It is very difficult to reach the larvæ under ground with any remedy other than digging for them and destroying them. Hogs are very fond of them, and, when turned into places where the grubs are abundant, will root up the ground and devour them in immense quantities. They are likewise eaten by domestic fowls and insectivorous birds; crows especially are so partial to them that they will often be seen following the plough, so as to pick out these choice morsels from the freshly-turned furrow. An insect parasite, the unadorned Tiphia, *Tiphia inornata* Say, is also actively engaged in destroying the white grub. Frequently, when digging the ground, a pale-brown, egg-shaped cocoon is turned up (see *c*, Fig. 220); within this, when fresh, will be found a whitish grub, represented at *b*, which, during its

growth, has fed upon the larva of the May-beetle. Within
this snug enclosure it soon changes to a pupa, and finally assumes the perfect form, as shown at *a* in the figure. The
fly is black, with sometimes a faint bluish tint, with dusky
wings, and the body more or less covered with pale-yellow
hairs, which are thickest on the under side.

A curious whitish fungus sometimes attacks this larva and
destroys it, growing out at the sides of the head; the protuberance or sprout rapidly increases in size, often attaining
a length of three or four inches, when it presents the appearance shown in Fig. 221. A very large number frequently
die from this cause. Trees infested with the beetles should
be shaken early in the morning, when the insects will fall, and
may be collected on sheets and killed by being thrown into
scalding water. Besides the cherry and plum, these insects
feed on the Lombardy poplar and the oaks. On account of
the length of time the larva takes to mature, the beetles are
not often abundant during two successive seasons.

No. 114.—The Cherry-tree Tortrix.

Cacœcia cerasivorana (Fitch).

Early in July there may often be found on the choke-
cherry, and sometimes also on the cultivated cherry, one or
more branches having all their leaves and twigs drawn
together with a web of silken threads. On opening one of
these enclosures, there will be found a large number of active
yellow larvæ. These are about five-eighths of an inch long,
nearly cylindrical, the head black, body above yellow, a little
paler between the segments, with a few very fine yellowish
hairs. The anterior portion of the second segment and the
hinder portion of the terminal one are black; there is also
a faint dorsal line of a darker shade. The under side is
similar to the upper in color, and the six anterior claw-like
feet are black.

The chrysalis is formed within the nest in which the larva
has lived, and is of a pale-brown color. The moth, when at

rest, is broad and flat, the outer edge of the fore wings being rounded towards the base, and straight from the middle to the tip, and when its wings are spread (see Fig. 222) it meas-

FIG. 222.

ures from three-quarters of an inch to an inch across. The fore wings are crossed by irregular wavy bands, alternately of bright ochre-yellow and pale, dull, leaden blue ; the yellow bands are varied with darker spots, the most conspicuous one of which is placed on the outer margin near the tip, and from this spot a broader ochre-yellow band extends towards the hind margin, and curves thence to the inner angle ; the hind wings and entire under surface are pale ochre-yellow.

Where this insect is found to be injurious, the webs containing the larvæ and chrysalids should be gathered and destroyed before the winged moths mature.

No. 115.—The Cherry-tree Plant-louse.

Myzus cerasi (Fabr.).

This black, disgusting-looking louse begins to appear on the leaves of the cultivated cherry almost as soon as they are expanded, being hatched from eggs deposited on the branches the previous autumn, and they multiply so fast that the under side of the young foliage is soon almost entirely covered with them, and the growth of the tree stunted by their continual appropriation of its juices. They crowd together in dense masses, often two deep, standing on each other's backs, with only sufficient space between to enable them to insert their extended beaks into the leaves. In a few days these insects multiply enormously, their black bodies covering not only the under side of the leaves but also the leaf-stalks, and clustering about the stems and green heads of the young fruit, while swarms of flies and other insects, attracted by the sweet exudations from the bodies of the lice, keep up a constant hum and buzz around the infested trees.

The presence of these aphides in such numbers has the

effect of attracting to the tree their natural enemies, which also multiply with great rapidity and make astonishing havoc among their defenceless victims. The lady-birds and their larvæ, also the larvæ of Syrphus flies and lace-wing flies, many of which are referred to under No. 57, appear in abundance among them, tearing and devouring them with the greatest ferocity, and usually within two or three weeks the armies of lice are completely annihilated, and the leaves of the trees appear clean again. Later in the season the lice appear a second time, but occupy only the tender leaves at the ends of the shoots, some of them usually remaining there during the rest of the summer. On the approach of cold weather, males are produced, and subsequently a stock of eggs is placed by the females about the base of the buds and in the fissures of the bark of the branches, where they remain unhatched until the following spring.

These lice may be killed by thoroughly drenching them with weak lye, strong soapsuds, or tobacco-water, but whatever solution may be used it must come in contact with the lice in order to be effectual; dipping the extremities of the limbs in such solutions, where such a course is practicable, will quickly destroy them. The easiest remedy, however, is to aid nature by introducing among the colonies a number of lady-birds and other enemies, who at once set to work to devour them with great vigor. A very minute Ichneumon fly, a species of Aphidius (*Trioxys cerasphis* Fitch), is parasitic upon these lice and destroys large numbers of them.

No. 116.—The Ursula Butterfly.

Limenitis ursula Fabr.

This is a medium-sized but handsome butterfly, which is seen on the wing during the months of June and July. It is represented in Fig. 223. Its wings are of a blackish-brown color glossed with a bluish tint, and with three marginal rows of bluish crescents of varying size. In the female the inner row is less marked, and each crescent is supported behind by a deep-orange patch or point. On the fore wings there are several white spots towards the tip. The margins of both wings are slightly crenate, the hollows being edged with white. When the wings are spread, they measure about three inches across.

The female deposits her eggs on the leaves of the cherry, both wild and cultivated, and occasionally also on those of

Fig. 223

the apple and plum. The full-grown larva is about an inch and a quarter long, of an olive-green color variegated with russet, white, reddish yellow, and ochreous, with two long reddish horns behind its head, and two tubercles on each of the other segments, all green except those on the fifth segment, which are reddish. The chrysalis is russety marked with white, is suspended by its tail, and has on the middle of its back a curious and prominent projection like a Roman nose. Both the larva and the chrysalis resemble that of *Limenitis disippus*, Fig. 180. This insect is met with only occasionally, and has never been reported as destructive anywhere. It is found as far north as the Province of Ontario in Canada, but is much more common in the Middle and Southern States.

No. 117.—The Cherry-tree Thecla.

Thecla titus Fabr.

This is a very pretty little butterfly, better known as *Thecla mopsus.* (See Fig. 224.) It is of a dark-brown color above, with a row of seven or eight orange-colored spots near the margin of the hind wings, which are larger and more conspicuous on the under than on the upper side. The wings beneath are light brown, with a row of deep but bright orange spots near the hind margins of both pairs, an inner and more irregular row of small black spots, encircled with white, and on the middle of the hind wings two similar spots, placed close together. In flight it is active, but its movements are of a jerky nature. The wings measure, when expanded, an inch and a quarter or more across.

Fig. 224.

The caterpillar, which is found feeding on cherry leaves during the month of May, is a curious flat creature, resembling a wood-louse in outline, of a dull-green color, pervaded by a yellowish tint. There is a patch of rose color on the anterior segments, and another larger one on the hinder extremity.

The chrysalis is pale brown and glossy, with many small dark-brown or blackish dots distributed over the whole surface, and thickly covered with very short brown hairs, scarcely visible without a magnifying-lens. The butterfly appears about the middle of July, and is very partial to the flowers of the "butterfly-weed," *Asclepias tuberosa,* as well as to those of the common milkweed, *Asclepias cornuti.*

This insect is never found in sufficient abundance to be injurious, but whenever met with it excites the curiosity of the observer.

ATTACKING THE FRUIT.

No. 118.—The Cherry Bug.

Metapodius femoratus (Fabr.).

Fig. 225.

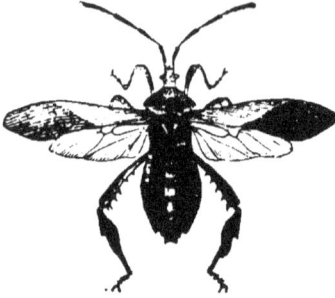

This insect, which belongs to the order *Hemiptera,* is said to injure the fruit of the cherry in the Western States by puncturing it with its beak and sucking the juices. It is represented in Fig. 225. It is said to attack only the sweet varieties of cherry.

SUPPLEMENTARY LIST OF INJURIOUS INSECTS WHICH AFFECT THE CHERRY.

ATTACKING THE ROOTS.

The larva of the stag-beetle, No. 5, also that of the rough Osmoderma, No. 8, occasionally injure the roots of the cherry, but chiefly affect those trees which are old and decaying.

ATTACKING THE BRANCHES.

The apple-twig borer, No. 13 ; the imbricated snout-beetle, No. 14; and the New York weevil, No. 100.

ATTACKING THE LEAVES.

The leaves of the cherry-tree suffer from all the following : the apple-tree tent-caterpillar, No. 20 ; the forest tent-caterpillar, No. 21 ; the white-marked tussock-moth, No. 22 ; the red-humped apple-tree caterpillar, No. 24 ; the canker-worms, Nos. 25 and 26 ; the fall web-worm, No. 27 ; the Cecropia emperor-moth, No. 28 ; the turnus swallow-tail, No. 30 ; the

American lappet-moth, No. 33; the oblique-banded leaf-roller, No. 35; the leaf-crumpler, No. 37; the eye-spotted bud-moth, No. 38; the many-dotted apple-worm, No. 43; the palmer-worm, No. 44; the hag-moth caterpillar, No. 48; the saddle-back caterpillar, No. 49; the tarnished plant-bug, No. 71; the pear-tree slug, No. 75; the gray dagger-moth, No. 84; the Disippus butterfly, No. 87; the blue-spangled peach-tree caterpillar, No. 102; the basket-worm, or bag-worm, No. 120; and the rose-beetle, No. 151. The pear-tree blister-beetle, No. 73, devours the blossoms as well as the young leaves.

ATTACKING THE FRUIT.

The plum curculio, No. 94, affects the fruit to an alarming extent in many sections, and, since the cherries do not drop from the trees as the plums do, from the injuries caused by this insect, the extent of its depredations is not easily ascertained. It is not unusual to find a considerable proportion of the ripe cherries in the markets containing the larva of this curculio, nearly full grown.

INSECTS INJURIOUS TO THE QUINCE.

ATTACKING THE TRUNK.

No. 119.—The Quince Scale.

Aspidiotus cydoniæ Comstock.

This scale is found on the quince-tree in Florida. It is of a gray color, somewhat transparent, very convex in form, and about six-hundredths of an inch in diameter. Where it is found injurious, it may be removed from the trunk and limbs with a stiff brush dipped in a strong solution of soap.

ATTACKING THE LEAVES.

No. 120.—The Basket-worm, or Bag-worm.

Thyridopteryx ephemeræformis (Haworth).

During the winter the curious weather-beaten bags of this insect may be seen hanging from many different sorts of trees, both evergreen and deciduous. In the latter class they are found on the quince, apple, pear, plum, cherry, peach, elm, maple, locust, and linden, and in the former on arbor-vitæ, Norway spruce, and red cedar. If a number of these bags are gathered in the winter and cut open, many of them will be found empty, but the greater portion will be seen to present the appearance shown at *e* in Fig. 226, being in fact partly full of soft, yellow eggs. Those which do not contain eggs are male bags, and the empty chrysalis skin of the male is generally found protruding from the lower end.

The eggs are soft, opaque, obovate in form, about one-twentieth of an inch long, and surrounded by more or less

222

fawn-colored silky down; they hatch during May or early
in June.

The young larvæ are of a brown color; they are very
active, and begin at once to make for themselves coverings of
silk, to which they fasten bits of the leaves of the tree on
which they are feeding, forming small cones, as shown at *g*
in the figure. As the larvæ grow, they increase the size of
their enclosures or bags from the bottom, until they become
so large and heavy that they hang instead of remaining

Fig. 226.

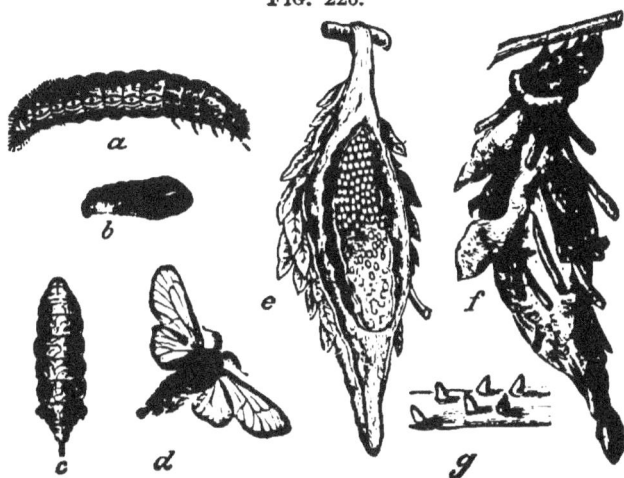

upright, as at first. By the end of July the caterpillars
become full grown, when they appear as shown at *f*, Fig.
226, where the larva is seen with its head and a portion of
its anterior segments protruded from the bag. When taken
out of the enclosure at this stage, it presents the appearance
shown at *a* in the figure, that portion of the body which has
been covered by the bag being soft, and of a dull-brownish
color, inclining to red at the sides, while the three anterior
segments, which are exposed when the insect is feeding or
travelling, are horny and mottled with black and white.
The small, fleshy prolegs on the middle and hinder segments
are fringed with numerous hooks, by which the larva is

enabled to cling to the silken lining of its bag and drag it along wherever it goes. The outer surface of the bag is rough and irregular from the protruding portions of the stems and leaves which are woven into it. During their growth these caterpillars are slow travellers, seldom leaving the tree on which they were hatched; but when about full grown they become much more active, and often lower themselves to the ground by silken threads, and slowly wander from place to place.

When about to change to chrysalids, they fasten their bags securely to the twigs of the trees on which they happen to be, and then undergo their change. The male chrysalis, shown at *b*, Fig. 226, is much smaller than the female, which is seen within the bag at *e*.

The female moth is wingless, and never leaves the bag, but works her way to its lower orifice, and there awaits the attendance of the male. She is not only without wings, but is destitute of legs also; in short, she seems to be nothing more than a yellowish bag of eggs with a ring of soft, pale-brown, silky hair near the tail. She is represented at *c* in the figure. The male (*d*, Fig. 226) has transparent wings and a black body, and is very active on the wing during the warmer portions of the day. After pairing, the female deposits her eggs, intermingled with fawn-colored down, within the empty pupa-case, and when this task is completed she works her way out of the case, drops exhausted to the ground, and dies.

The bag-worm is a Southern rather than a Northern insect, although it is found as far north as New Jersey and New York, and occasionally in Massachusetts; it is extremely local in its character, often abounding in one particular neighborhood and totally unknown a few miles away. Where they occur in abundance they often almost entirely defoliate the trees they attack; this, however, may be easily prevented by gathering the cases which contain the eggs for the next brood during the winter and destroying them. There are two species of Ichneumon which attack the bag-worm: one of them, *Cryptus inquisitor* (Say) (Fig. 227), is about two-fifths of an inch long, the other, *Hemiteles thyridopteryx* Riley, is about one-third of an inch long; the male is shown in Fig. 228, the female in Fig. 229, both magnified. Five or six of this

FIG. 227.	FIG. 228.	FIG. 229.	FIG. 230.

latter species will sometimes occupy the body of a single caterpillar. After destroying their victim they spin for themselves tough, white, silken cocoons within the bag, a section of which is shown in Fig. 230.

ATTACKING THE FRUIT.

No. 121.—The Quince Curculio.

Conotrachelus cratægi Walsh.

This is a broad-shouldered snout-beetle, larger than the plum curculio, No. 94, and has a longer snout; in Fig. 231, *a* shows a side view of the insect, *b* a back view. It is of an ash-gray color, mottled with ochre-yellow and whitish, with a dusky almost triangular spot at the base of the thorax above, and seven narrow longitudinal elevations on the wing-covers, with two rows of dots between each. It is an indigenous insect, having its home in the wild haws, in which it is frequently found, but it is also very injurious to the quince. It appears during the month of June, and punctures the young fruit, making a cylindrical

FIG. 231.

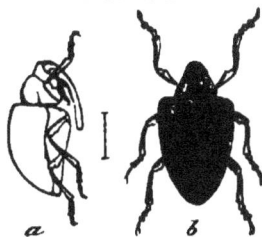

15

hole a little larger than is sufficient to admit the egg, and enlarged at the base. Within this receptacle the egg is placed, and hatches there in a few days. The larva does not penetrate to the core, but burrows in the fruit near the surface; it resembles the larva of the plum curculio in appearance, but is somewhat larger, and has a narrow dusky line down the back. In about a month it becomes full grown, when it leaves the fruit through a cylindrical opening and buries itself two or three inches in the ground, where it remains during the autumn, winter, and early spring months without change. It becomes a pupa early in May, and assumes the beetle form a few days afterwards. The beetle also feeds on the quince, burying itself completely in the substance of the fruit; it occasionally attacks the pear.

Where these beetles prove destructive they may be collected by jarring, as recommended for the plum curculio; and care should be taken to destroy all the fruit which falls prematurely to the ground.

SUPPLEMENTARY LIST OF INJURIOUS INSEOTS WHIOH AFFEOT THE QUINOE.

ATTACKING THE TRUNK.

The round-headed apple-tree borer, No. 2.

ATTACKING THE LEAVES.

The leaf-crumpler, No. 37; the tarnished plant-bug, No. 71; and the pear-tree slug, No. 75. The pear-tree blister-beetle, No. 73, eats both the flowers and the leaves.

ATTACKING THE ROOTS.

No. 122.—The Broad-necked Prionus.

Prionus laticollis (Drury).

This is a gigantic borer (Fig. 232), from two and a half to three inches in length, of a yellowish-white color, with a

FIG. 232.

small, horny, reddish-brown head, and a bluish line down the back, which cuts for itself a cylindrical hole through the centre of the root of the vine, a little below the surface; and when the root is barely large enough to contain the larva, nothing but a thin skin of bark is left, but this is always found entire, so that the insect cannot be easily discovered. It is probable that it lives in the larval state about three years, and that it changes to a pupa (Fig. 233) within the root towards the end of June.

The beetle appears about the middle of July, and is known as the Broad-necked Prionus. Fig. 234 represents the female, which measures from an inch and a quarter to an inch and three-quarters in length, and is of a brownish-black color, with strong, thick jaws; the antennæ are rather slender; the thorax is short and wide and armed at the sides with three teeth. The wing-covers have three slightly-elevated lines on each, and are thickly punctated. In the male the body is shorter, while the antennæ are longer, stouter, and toothed.

Little or nothing can be done in the way of extirpating these under-ground borers, as their presence is seldom suspected

FIG. 233.

FIG. 234.

until the vine becomes sickly, or dies from the injuries they have caused. Where grape-vines die suddenly from any unknown cause, the roots should be carefully examined, and if evidences of the presence of this borer are discovered, it should be searched for and destroyed.

No. 123.—The Tile-horned Prionus.

Prionus imbricornis (Linn.).

The larva of this beetle, a species closely allied to No. 122, has also been found devouring the roots of the grape-vine. The larvæ of these two species resemble each other so closely that they are almost indistinguishable. When full grown, the borer collects together a few fibres and chips of the roots, and with the aid of these constructs a loose cocoon, within which it changes to a pupa almost identical with that of No. 122. (See Fig. 233.)

This beetle, which is represented in Fig. 235, is called the Tile-horned Prionus because the joints of the antennæ of the male overlap one another like tiles on a roof. It is very similar in appearance to the broad-necked prionus, but the two species may be distinguished by the difference in the

Fig. 235.

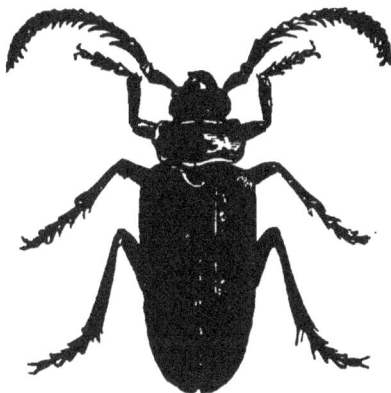

number of the joints in their antennæ: in *imbricornis* the male has about nineteen joints, and the female about sixteen, while in *laticollis* both sexes have twelve-jointed antennæ. Any remedial measures useful for one species will be equally applicable to the other.

No. 124.—The Grape-vine Root-borer.

Ægeria polistiformis Harris.

This larva resembles that of the peach-tree borer, No. 97, in appearance and habits, but is a little larger in size. The larvæ of the Prionus beetles have only six legs, while this Egerian larva, in common with most lepidopterous insects, has sixteen legs,—six horny ones on the anterior segments, and ten fleshy or membranous ones on the hinder segments,—and when full grown it measures from an inch to

Fig. 236

an inch and a half in length. (See Fig. 236.) It lives exclusively under ground, and consumes the bark and sap-wood of the grape-roots, eating irregular furrows into their substance; sometimes it eats the bark, and at other times works its way under the surface.

When full grown, the larva forms a pod-like cocoon of a gummy sort of silk, covered with little bits of wood, bark, and earth, and situated within or adjacent to the injured root. Within this it changes to a brown chrysalis, which, when mature, works itself out of the cocoon by means of minute

Fig. 237.

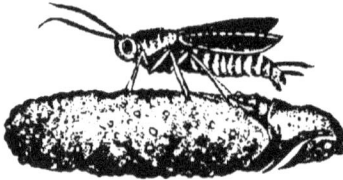

teeth, with which the segments are armed, and thence to the surface of the ground, when the perfect insect escapes. Fig. 237 shows the cocoon with the chrysalis partly protruding from it and the newly-escaped moth resting on it.

The moth resembles a wasp in appearance, and in the noise it makes during its flight. The female is shown in Fig. 238. The antennæ are simple and black, the body of a brownish-black color, marked with orange or tawny yellow. There is a bright-yellow band on the base of the second segment of its abdomen, and usually a second one on the fourth joint, but sometimes this latter is wanting; near the tip of the abdomen below there is a short pencil of tawny orange hairs on each side. The fore wings are brownish black, with a more or less distinct clear patch at the base; the hind wings transparent, with the veins, the terminal edge, and the fringe brownish black. In the male (Fig. 239) the antennæ are toothed, except for a short distance near the tip; the thorax and abdomen are darker in color, and in addition to the short pencils of orange hairs on the abdomen below, there are two longer ones above. The wings, when expanded, measure from an inch to an inch and a half across. The moth appears during August.

The female is said to deposit her eggs on the collar of the grape-vine, close to the earth, and the young larvæ, as soon as hatched, descend to the roots.

This insect inhabits the Middle, Western, and some of the Southern States. It is said to have been exceedingly destructive in North Carolina both to wild and cultivated grapes, and is reported as injurious also in Kentucky. The moth is found in the South from the latter part of June until September.

It is stated that the Scuppernong grape, a variety of the fox-grape, *Vitis vulpina,* is never attacked by this borer; if this

FIG. 238.

FIG. 239.

be so, its ravages may be prevented by grafting other vines on roots of the Scuppernong. When it has been ascertained that the borers are at work on a vine, the earth should be cleared away from above the roots and the invaders searched for and destroyed; hot water applied about the roots is said to kill them. As a preventive measure, mounding the vines, as recommended for peach-trees, under the head of the peach-tree borer, No. 97, would probably be beneficial.

No. 125.—The Grape Phylloxera.

Phylloxera vastatrix Planchon.

This tiny foe to the grape-vine has attained great celebrity during the past few years, and much attention has been paid to the study of its life-history and habits, in the hope of devising some practical measures for its extermination. The destruction it has occasioned in France has been so great that it has become a national calamity, which the government has appointed special agents to inquire into; large sums of money have also been offered as prizes to be given to any one who shall discover an efficient remedy for this insect pest. At the same time it has made alarming progress in Portugal, also in Switzerland and in some parts of Germany, and among vines under glass in England. It is a native of America, whence it has doubtless been carried to France; it is common throughout the greater portion of the United States and in one of its forms in Canada; but our native grape-vines seem to endure the attacks of the insect much better than do those of Europe. Recently it has appeared on the Pacific slope, in the fertile vineyards of California, where the European varieties are largely cultivated, and hence its introduction there will probably prove disastrous to grape-culture.

This insect is found in two different forms: in one instance on the leaf, where it produces greenish-red or yellow galls of various shapes and sizes, and is known as the type *Gallæcola*, or gall-inhabiting; in the other and more destructive form, on the root, known as the type *Radicicola*, or root-inhabiting, causing at first swellings on the young rootlets, followed by decay, which gradually extends to the larger roots as the insects congregate upon them. These two forms will for convenience be treated together, a slight departure from the general plan of this work.

The first reference made to the gall-producing form was by Dr. Fitch in 1854, in the "Transactions of the New York State Agricultural Society," where he described it under the name of *Pemphigus vitifoliæ*. Early in June there appear upon the vine leaves small globular or cup-shaped galls of varying sizes. A section of one of these is shown at *d*, Fig. 241; they are of a greenish-red or yellow color, with their outer surface somewhat uneven and woolly. Fig. 240 represents a leaf badly infested with these galls. On opening one of the freshly-formed galls, it will be found to contain from one to four orange-colored lice, many very minute, shining, oval, whitish eggs, and usually a considerable number of young lice, not much larger than the eggs, and of the same color. Soon the gall becomes over-populated, and the surplus lice wander off through its partly-opened mouth on the upper side of the leaf, and establish themselves either on the same leaf or on adjoining young leaves, where the irritation occasioned by their punctures causes the formation of new galls, within which the lice remain. After a time the older lice die, and the galls which they have inhabited open out and gradually become flattened and almost obliterated; hence it may happen that the galls on the older leaves on a vine will

Fig. 240.

be empty, while those on the younger ones are swarming with occupants.

These galls are very common on the Clinton grape and other varieties of the same type, and are also found to a greater or less extent on most other cultivated sorts. They sometimes occur in such abundance as to cause the leaves to turn brown and drop to the ground; and instances are recorded where vines have been defoliated from this cause. The number of eggs in a single gall will vary from fifty to four or five hundred, according to its size. There are several generations of the lice during the season, and they continue to extend the sphere of their operations throughout the greater part of the summer. Late in the season, as the leaves become less succulent, the lice seek other quarters, and many of them find their way to the roots of the vines and establish themselves on the smaller rootlets. By the end of September the galls are usually deserted. In Fig. 241 we have this type

FIG. 241.

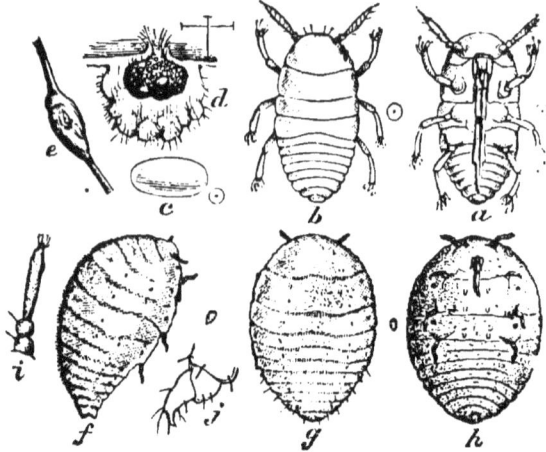

of the insect illustrated : *a* shows a front view of the young louse, and *b* a back view of the same, *c* the egg, *d* a section of one of the galls, *e* a swollen tendril, *f, g, h,* mature egg-bearing gall-lice, lateral, dorsal, and ventral views, *i* antennæ, and *j* the two-jointed tarsus.

When on the roots, the lice subsist also by suction, and their punctures result in abnormal swellings on the young rootlets, as shown at *a* in Fig. 242. These eventually decay, and this decay is not confined to the swollen portions, but involves the adjacent tissue, and thus the insects are induced to betake themselves to fresh portions of the living roots, until at last the larger ones become involved, and they, too, literally waste away.

In Fig. 242 we have the root-inhabiting type, *Radicicola,* illustrated : *a,* roots of Clinton vine, showing swellings; *b,* young louse, as it appears when hibernating ; *c, d,* antennæ and leg of same ; *e, f, g,* represent the more mature lice.

It is also further illustrated in Fig. 243, where *a* shows a healthy root, *b* one on which the lice are working, *c* a root which is decaying and has been deserted by them ; *d, d, d,* indicate how the lice are found on the larger roots; *e* represents the female pupa, seen from above, *f* the same from below, *g* winged female, dorsal view, *h* the same, ventral view, *i* the antennæ of the winged insect, and *j* the wingless female, laying eggs on the roots; *k* indicates how the punctures of the lice cause the larger roots to rot. Most of these figures are

FIG. 242.

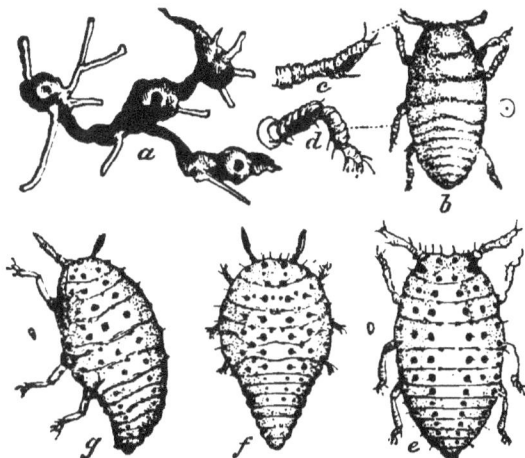

highly magnified, the short lines or dots at the side showing the natural size.

During the first year of the insect's presence the outward manifestations of the disease are very slight, although the fibrous roots may at this time be covered with the little swellings; but, if the attack is severe, the second year the leaves assume a sickly yellowish cast, and the usual vigorous yearly growth of cane is much reduced. In course of time the vine usually dies; but, before this takes place, the lice, having little or no healthy tissue to work on, leave the dying vine and seek for food elsewhere, either wandering under ground among the interlacing roots of adjacent vines, or crawling over the

surface of the ground in search of more congenial quarters.
During the winter many of them remain torpid, and at that
season they assume a dull-brownish color, so like that of the

FIG. 243.

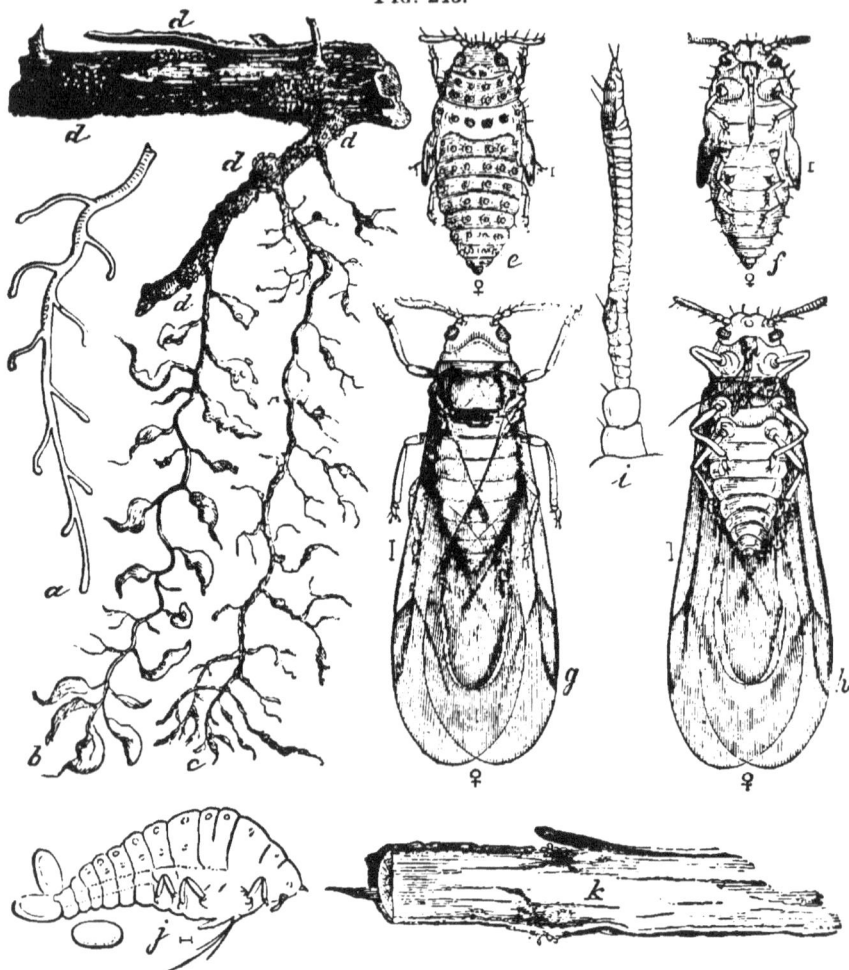

roots to which they are attached that they are difficult to
discover. They have then the appearance shown at *b* in Fig.
242. With the renewal of growth in the spring, the young
lice cast their coats, rapidly increase in size, and appear as
shown at *e, f, g,* in the figure; soon they begin to deposit eggs;

these eggs hatch, and the young ones shortly become egg-laying mothers like the first, and, like them, also remain wingless. After several generations of these egg-bearing lice have been · produced, a number of individuals about the middle of summer acquire wings. These also are all females, and they issue from the ground, and, rising in the air, fly, or are carried with the wind, to neighboring vineyards, where they deposit eggs on the under side of the leaves among their downy hairs, beneath the loosened bark of the branches and trunk, or in crevices of the ground about the base of the vine. Occasionally individual root-lice abandon their underground habits and form galls on the leaves.

The complete life-history of this insect is extremely interesting and curious, and those desiring further information as to the different modifications of form assumed by the insect in the course of its development will find it given with much minuteness of detail in the fifth, sixth, seventh, and eighth " Reports on the Insects of Missouri," by C. V. Riley.

Remedies.—This is an extremely difficult insect to subdue, and various means for the purpose have been suggested, none of which appear to be entirely satisfactory. Flooding the vineyards, where practicable, seems to be more successful than any other measure, but the submergence must be total and prolonged to the extent of from twenty-five to thirty days; it should be undertaken in September or October, when it is said that the root-lice will be drowned and the vines come out uninjured.

Bisulphide of carbon is stated by some to be an efficient remedy; it is introduced into the soil by means of an auger with a hollow shank, into which this liquid is poured; several holes are made about each vine, and two or three ounces are poured into each hole. Being extremely offensive in odor and very volatile, its vapor permeates the soil in every direction, and is said to kill the lice without injuring the vines. This substance should be handled with caution, as its vapor is very inflammable and explosive. Alkaline sulpho-carbonates are also recommended ; these are gradually decomposed in the soil and give off sulphuretted hydrogen and bisulphide of carbon.   Carbolic acid mixed with water, in the proportion of one part of the acid to fifty or one hundred parts of water, has also been used with advantage, poured into two or three holes made around the base of each vine with an iron bar to the depth of a foot or more.   Soot is also recommended to be strewed around the vines.

It is stated that the insect is less injurious to vines grown on sandy soil, also to those grown on lands impregnated with salt.

Since large numbers of these insects, both winged and wingless, are known to crawl over the surface of the ground in August and September, it has been suggested to sprinkle the ground about the vines at this period with quicklime, ashes, sulphur, salt, or other substances destructive to insect life.   The application of fertilizers rich in potash and ammonia, such as ashes mixed with stable-manure or sal ammoniac,- has been found useful.   A simple remedy for the gall-inhabiting type is to pluck the leaves as soon as the galls appear and destroy them.

Several species of predaceous insects prey on this louse. A black species of Thrips with white-fringed wings (*Thrips phylloxeræ* Riley, see Fig. 244) deposits its eggs within the

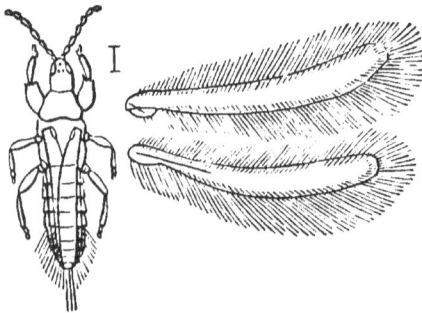
Fig. 244.

gall, which when hatched produce larvæ of a blood-red color, which play sad havoc among the lice. The larva of a Syrphus fly, *Pipiza radicum*, which feeds on the root-louse of the apple (see Fig. 2), has also been found attacking the Phylloxera.   Another useful friend is a small mite (*Tyroglyphus phylloxeræ* P. & R., see Fig. 245), which devours the lice; and associated with this is sometimes found another species (*Hoplophora arctata* Riley) of a very curious form, reminding one of a mussel. Fig. 246 represents this insect in different attitudes, highly magnified.

The gall-inhabiting type is very subject to the attacks of a small two-winged fly, *Diplosis grassator* Fyles, which deposits

Fig. 245.

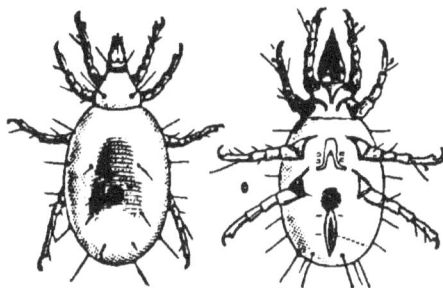

its eggs either in the gall or at its entrance, from which the larva is soon produced. This, although destitute of legs, is very active, and, groping about in the interior of the gall, seizes on the young lice soon after they are hatched and sucks them dry. It does not appear at first to attack the parent lice;

Fig. 246.

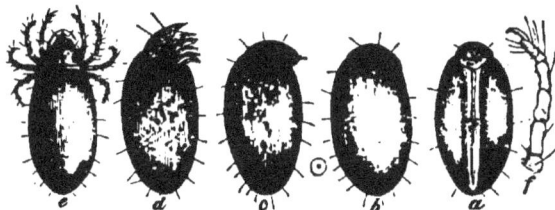

the tender progeny are more to its liking, and these are produced in sufficient numbers to furnish it with a constant supply of fresh food. In some instances one larva, in others two are found in a single gall, and as they increase in size they devour the lice very rapidly, and before changing to the

pupa state clear the gall entirely of its contents. The larva (Fig. 247, *a*) is about one-tenth of an inch long, of a pale piukish-yellow color, glossy and semi-transparent, with a dark line down the back on the two anterior and some of the posterior segments. On the terminal segment there are two short, fleshy horns united by a slight ridge; the horns are tipped with brownish black, and have a minute cluster of spines at their summit.

FIG. 247.

The pupa shown at *b* in the figure, is a little less than one-tenth of an inch in length, of a reddish-brown color, with a few short hairs scattered over its surface, and two blackish horns united by a ridge near the hinder extremity. Both the pupa and the larva are magnified.

The perfect insect escapes in about a fortnight after the pupa is formed. It is a very pretty little two-winged fly, shown much magnified at *c* in the figure, and of its natural size at *d*.

The Phylloxera is also preyed on by the larva of a dull-colored lady-bird, a species of Scymnus, by several other species of the lady-bird family, and by the larvæ of the lace-wing flies referred to under No. 57.

To guard against its introduction into new vineyards, the roots of young vines should be carefully examined before being planted, and if knots and lice are found upon them these latter may be destroyed by immersing the roots in hot soap-suds or tobacco-water.

Our native American vines are found to withstand the attacks of this insect much better than do those of European origin; hence by grafting the more susceptible varieties on these hardier sorts, the ill effects produced by the lice may in some measure be counteracted. The roots recommended to be used as stocks are those of Concord, Clinton, Herbemont, Cunningham, Norton's Virginia, Rentz, Cynthiana, and Taylor. The Clinton, one of the varieties recommended, is particularly liable to the attacks of the gall-producing type of Phylloxera, but the lice are seldom found to any great extent on its roots, and the vine is so vigorous a grower that a slight attack would not produce any perceptible injury.

ATTACKING THE BRANCHES.

No. 126.—The Grape-vine Bark-louse.

Pulvinaria innumerabilis Rathvon.

During the month of June there are sometimes found on the branches of the grape-vine, brown, hemispherical scales, from under one end of which there protrudes a cotton-like substance, which increases in size until the beginning of July, by which time it has become a mass about four times as large as the scale. (See Fig. 248.) This cottony matter contains the eggs of the insect, and very soon there issue from it minute, oval, yellowish-white lice, which distribute themselves over the branches, to which they attach themselves, and shortly become stationary, sucking the juices. This species is believed to be the same as the European scale-insect of the vine. These scales are not usually found in any great abundance, and may be readily scraped off with a knife or other suitable instrument, which should be done before the young lice escape.

Fig. 248.

16

No. 127.—The Four-spotted Spittle-insect.

Aphrophora 4-notata Say.

Occasionally there appear upon the branches in June spots of white, frothy matter, resembling spittle, embedded in which is found a soft, pale, wingless insect, which punctures the bark and sucks the juices from the branch, at the same time secreting over and around itself this spittle-like covering. The perfect or winged insect (see Fig. 249) is a flattened tree-hopper of a brown color, which occurs upon the vines in the early

Fig. 249. part of July. It is about three-tenths of an inch long; its wing-covers are brown, with a blackish spot at the tip, a second one on the middle of the outer margin, and a third one at the base, with the spaces between the spots whitish. Should this insect at any time prove injurious, it may be easily destroyed by the hand while in the soft, wingless form enclosed in its frothy covering.

No. 128.—Signoret's Spittle-insect.

Aphrophora Signoreti Fitch.

This is an insect very similar in habits and appearance to No. 127, surrounding itself while in the soft or larval condition with the same sort of frothy mass. When perfect, it is a little more than three-tenths of an inch long, of a tawny-brown color clouded with dull white, and thickly punctated with black dots. The wing-covers have on their inner margin, near the tip, a small white spot, and another larger one opposite this on the outer margin; but the wings are not spotted with black as in No. 127.

No. 129.—The Two-spotted Tree-hopper.

Enchenopa binotata (Say).

This is a small but very odd-looking brown insect, with two yellowish spots on the edge of the back, and a prolongation in front like the beak of a bird. It sometimes punctures the tender stems of the grape, causing them to wilt and turn brown. While this tree-hopper is occasionally found on the vine, it is much more common on the red-bud, *Cercis;* but its favorite home is on the wafer-ash, *Ptelea trifoliata.*

No. 130.—The Red-shouldered Sinoxylon.
Sinoxylon basilare (Say).

The larva of this insect (Fig. 250, *a*) bores into the stems of grape-vines, and sometimes also into the branches and trunks of apple and peach trees. It is a yellowish, wrinkled grub, about three-tenths of an inch long, with the anterior segments swollen, the head small, and the body arched or bent.

FIG. 250.

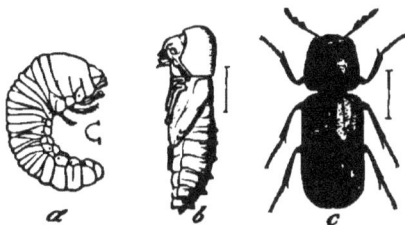

The pupa (Fig. 250, *b*) is of a pale-yellowish color, and is formed in the chambers mined by the larva.

The beetle is shown in the figure at *c*. It is about one-fifth of an inch long, black, with a large reddish spot at the base of each wing-cover. The thorax is punctated and armed with short spines in front; the wing-covers are roughened with dots, and appear as if cut off obliquely behind, the outer edge of the cut portion being furnished with three teeth on each side.

The only method suggested for destroying this insect is to burn the wood infested by it.

No. 131.—The Grape-vine Wound-gall.
Vitis vulnus Riley.

This curious gall, which is represented in Fig. 251, is produced by the Sesostris snout-beetle, *Ampeloglypter Sesostris* (Lec.). The beetle (Fig. 252) is about one-eighth of an inch long, of a reddish-brown color, with a stout beak half as long as its body. The thorax is punctated, and the wing-cases are polished and glossy, without any markings. It appears during the early part of July, when the female punctures the stem of the vine and deposits an egg therein, which shortly hatches, producing a tiny whitish grub, which lives within the swollen part and feeds upon it. At first the gall is small and inconspicuous, but towards the end of the season it assumes the form of an elongated knot or swelling, as shown in the figure; this is generally situated immediately above or below a joint. Usually there is a longitudinal depression on one side, dividing that portion into two prominences, which commonly have a rosy tint. Within the gall the larva remains until June of the following year. When full grown, it is about a quarter of an inch long, white, cylindrical, and footless, with a large yellowish head. During the month of June it changes to a pupa, from which the perfect beetle is produced in about a fortnight.

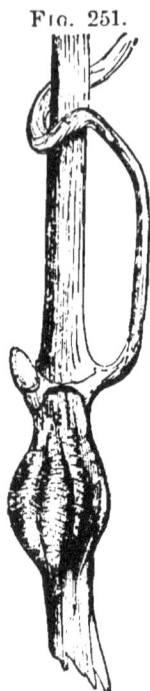

Fig. 251.

Fig. 252.

These galls do not appear to injure to any material extent the branches on which they occur; should they ever multiply so as to become injurious, their increase may be readily checked by cutting off and burning those portions of the canes on which they are situated, before the beetles escape.

ATTACKING THE LEAVES.

No. 132.—The Green Grape-vine Sphinx.

Darapsa myron (Cramer).

The larva of this insect is one of the most common and destructive of the leaf-eating insects injurious to the grape.

The first brood of the perfect or winged insect appears from the middle to the end of May, when the female deposits her eggs on the under side of the leaves, generally placing them singly, but sometimes in groups of two or three. The eggs are nearly round, about one-twentieth of an inch long, a little less in width, smooth, and of a pale yellowish-green color, changing to reddish before hatching.

The young caterpillar comes out of the egg in five or six days, when it makes its first meal on a part of the empty egg-

Fig. 253.

shell, and then attacks the softer portions of the grape-vine leaves. When first hatched, it is one-fifth of an inch long, of a pale yellowish-green color, with a large head, and having a long black horn near its posterior extremity, half as long as its body. As it increases in size, the horn becomes relatively shorter and changes in color; the markings of the larva also vary considerably at each moult. When full grown, it presents the appearance shown in Fig. 253. It is then about two inches long, with a rather small head of a pale-green color dotted with yellow and with a pale-yellow stripe down each side; the body is green, of a slightly deeper shade than the head, and covered with small yellow dots or granulations;

along the sides of the body these granulations are so arranged as to form a series of seven oblique stripes, extending backwards, and margined behind with a darker green. A white lateral stripe with a dark-green margin extends from just behind the head to the horn near the other extremity. Along the back are a series of seven spots, varying in color from red to pale lilac, each set in a patch of pale yellow. The caudal horn is one-fifth of an inch long, and varies in color from reddish to bluish green, granulated with black in front, and sometimes yellow behind and at the tip. This larva has the power of drawing the head and next two segments within the fourth and fifth, causing these latter to appear much distended; the feet are red, the prolegs pale green. Some specimens, especially among those of the later brood, will be found exhibiting remarkable variations in color; instead of green they assume a delicate reddish-pink hue, with markings of darker shades of red and brown, which so alter their appearance that they might at first sight be readily taken for a different species; a careful comparison, however, will show the same arrangement of dots and spots as in the normal form.

When full grown, the larva descends from the vine and draws a few leaves loosely together, binding them with silken threads, usually about or near the base of the vine on which

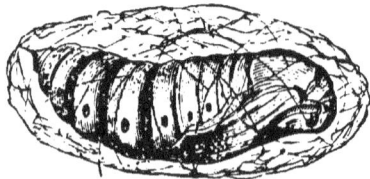

FIG. 254.

it has fed, and within this rude structure changes to a chrysalis (see Fig. 254) of a pale-brown color, dotted and streaked with a darker shade, and with a row of oval dark-brown spots along each side.

The moths from this first brood of larvæ usually appear during the latter part of July, when they deposit eggs for a second brood, which mature late in September, pass the winter in the pupa state, and emerge as moths in the following May.

The wings of this insect, when fully expanded, measure

about two and a half inches across, their form being long and narrow, as shown in Fig. 255. The fore wings are of a dark olive-green color, crossed by bands and streaks of greenish gray, and shaded on the outer margin with the same hue. The hind wings are dull red, with a patch of greenish gray next the body, shading gradually into the surrounding color. On the under side the red appears on the fore wings, the hinder pair being greenish gray. The antennæ are dull white above, rosy below, head and shoulder-covers deep olive-green, the

Fig. 255.

rest of the body of a paler shade of green; underneath the body is dull gray.

This moth rests quietly during the day, taking wing at dusk, when it is extremely active ; its flight is very swift and strong, and as it darts suddenly from flower to flower, rapidly vibrating its wings, remaining poised in the air over the objects of its search, while the long, slender tongue is inserted and the sweets extracted, it reminds one strongly of a humming-bird.

The caterpillars are very destructive to the foliage of the vine, being capable of consuming an enormous quantity of food ; one or two of them, when nearly full grown, will almost strip a small vine of its foliage in the course of two or three days. In some districts they are said to nip off the stalks of the half-grown clusters of grapes, so that they fall unripe to the ground.

Remedies.—The readiest and most effectual method of disposing of these pests is to pick them off the vines and kill them. They are easily found by the denuded canes which mark their course, or where the foliage is dense they may be tracked by their large brown castings, which strew the ground under their places of resort. Nature has provided a very efficient check to their undue increase, in a small parasitic fly, a species of Ichneumon (see Fig. 256), the female of which punctures the skin of the caterpillar and deposits her eggs underneath, where they soon hatch into young larvæ, which feed upon the fatty portions of their victim, avoiding the vital organs. By the time the sphinx caterpillar has become full grown, these parasitic larvæ have matured, and, eating their way through the skin of their host, they construct their tiny snow-white cocoons on its body, as shown in Fig. 257, from which, in about a week, the friendly fly escapes by pushing open a nicely-fitting lid at one end of its structure. No larva thus infested ever reaches maturity; it invariably shrivels up and dies.

Fig. 256.

Fig 257.

No. 133.—The Pandorus Sphinx.

Philampelus Pandorus (Hübn.).

This is one of the most beautiful of our Sphinx moths, a rare as well as lovely creature, and an object highly prized by collectors. It is found throughout the Northern United States, and occasionally in Canada, but is nowhere very common. It is represented in Fig. 258. Its wings, when expanded, will measure from four to four and a half inches across; they are of a light-olive color, mixed with gray, and varied with patches of a darker olive-green, rich and velvety, and some portions, especially on the hind wings, of a rosy hue. The body is pale greenish brown, ornamented with dark-olive patches. The moths appear in July, when, after pairing, the female deposits her eggs singly on the leaves of the grape-vine, or Virginia creeper, *Ampelopsis quinquefolia*, where they shortly hatch, producing small green larvæ of a pinkish hue along the sides, and with a very long pink horn at the tail. As the caterpillar increases in size, the horn becomes shorter,

Fig. 258.

. and after a time curves round, as shown at *c*, Fig. 259. As the larva approaches maturity, it changes to a reddish-brown color, and after the third moult entirely loses the caudal horn, which is replaced by a glassy, eye-like spot. The mature larva, when in motion, as shown at *a* in the figure, will measure nearly four inches in length, but when at rest it draws the head and two adjoining segments within the fourth, as shown in the figure at *b*, which shortens its body nearly an inch, giving it a very odd appearance, with its anterior portions so blunt and thick. It is of a rich reddish-brown color, of a lighter shade along the back, with five nearly oval cream-colored spots along each side from the seventh to the eleventh segment inclusive. On the anterior segments there are a number of black dots; a dark, polished, raised, eye-like spot in place of the tail, the breathing-pores along the sides black, showing prominently in the cream-colored spots. It is a very voracious feeder, and strips the vine of its leaves with such rapidity that it soon attracts attention.

Fig. 259.

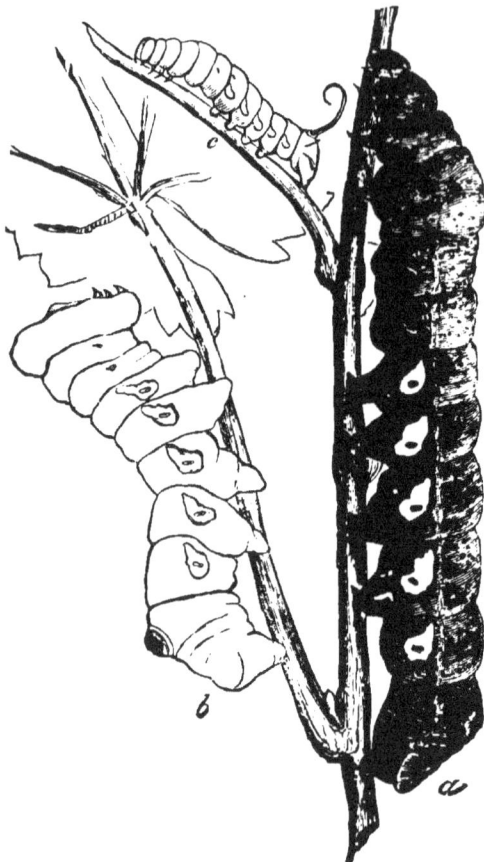

When full grown, it descends from the vine and buries itself in the ground, where it forms an oval cell, within which it changes to a chrysalis. The chrysalis is of a chestnut-brown color, with the segments roughened with impressed points, the terminal joint having a long thick spine. The insect usually remains in the chrysalis state until the following summer, but occasionally it matures and escapes the same season. Should these larvæ at any time prove troublesome, they can be readily subdued by hand-picking.

No. 134.—The Achemon Sphinx.

Philampelus achemon (Drury).

The caterpillar of this sphinx (Fig. 260) is truly a formidable-looking creature, measuring, when full grown, if at rest, about three inches, and when in motion about three and a half inches. It much resembles that of Pandorus, No. 133, and feeds also on the Virginia creeper (*Ampelopsis quinquefolia*) as well as on the grape-vine. The egg is laid on the under side of the leaf in July, and the young larva, when hatched, is of a light-green color, with a very conspicuous reddish-brown horn, half as long as its body, which, as the larva increases in size, becomes shorter, and finally disappears, its place being occupied by a polished tubercle with a central black dot. The mature larva varies from a pale straw-color to a reddish brown, the color growing darker down the sides,

Fig. 260.

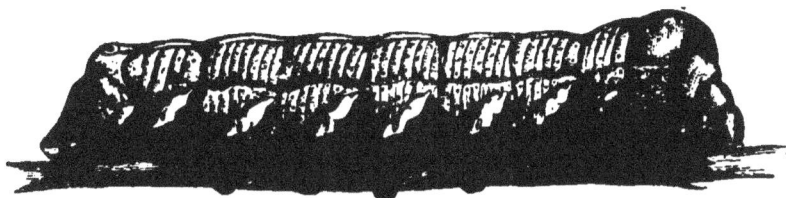

becoming deep brown as it approaches the under surface. An interrupted line of brown runs along the back, and another unbroken one extends along each side ; below this latter there are six cream-colored spots, as shown in the figure, one on each segment, from the sixth to the eleventh inclusive. The body is much wrinkled, and dotted with minute spots, which are dark on the back, lighter and annulated at the sides. The head and next two segments are small, and are drawn within the fourth when at rest, as seen in the figure. It becomes full grown during the latter part of August or early in September, and just before undergoing its next change assumes a beautiful pink or crimson color.

Leaving the vine, it descends to the ground, where it buries itself to the depth of several inches, and, having formed for itself a smooth cell, changes to a chrysalis (Fig. 261) of a dark, shining, mahogany color, with the anterior edges of the segments along the back roughened with minute points, and with a short, blunt spine at the extremity. The insect usually remains in this condition in the ground until late in June the following year; but instances have been recorded where the moth has appeared the same season.

FIG. 261.

FIG. 261.

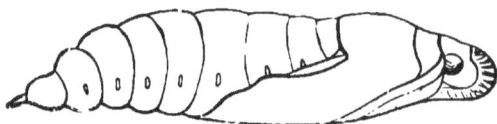

The moth is of a brownish-gray color, variegated with light brown, and with deep-brown spots, as shown in Fig. 262. The hind wings are pink, becoming deeper red near the middle. There is a broad gray border behind, with a row of darker

FIG. 262.

spots along its front edge, becoming fainter towards the outer margin. The body is reddish gray, with two triangular patches of deep brown on the thorax.

This insect is found in almost all parts of the United States and Canada where the grape is cultivated, but has never occurred in sufficient numbers to be injurious. It is so conspic-

uous in the larval state that it might easily be controlled by
hand-picking should it at any time prove troublesome.

No. 135.—The Abbot Sphinx.

Thyreus Abbotii Swainson.

This is not a common insect, yet it is found occasionally
over a large portion of the United States and Canada. The
caterpillar (see Fig. 263) attains full growth about the end of

FIG. 263.

July or the beginning of August, when it measures nearly two
and a half inches in length. It varies considerably in color,
from dull yellow to reddish brown, each segment being marked
transversely with six or seven fine black lines, and longitudinally with dark-brown patches, giving to the larva a checkered appearance. Near the posterior extremity of the body
there is a polished black tubercle above, ringed with yellow.

The chrysalis is commonly formed in a little cavity on the
surface of the ground, covered with a few pieces of leaves
loosely fastened together and mixed with grains of earth, but
it is said sometimes to bury itself below the surface. It is
about an inch and a quarter long, of a dark-brown color,
roughened with small indentations except between the joints,

and terminating in a flattened point, with two small thorns at the end. The insect remains in the chrysalis condition until the following spring.

The moth (Fig. 263) is found on the wing from the early part of April to the end of May, and measures, when its wings are spread, two and a half inches or more across. It is of a dull chocolate-brown color, the front wings becoming pale beyond the middle, and marked with dark brown as in the figure. The hind wings are yellow, with a broad brown border, breaking into a series of short lines as it approaches the body. The abdomen is furnished with tufts along the sides near the extremity, and when the insect is at rest is curved upwards.

It is scarcely likely that it will ever prove destructive; should it at any time become so, it may be subdued by hand-picking. It is preyed upon by a small species of Ichneumon fly, which in the larval state lives within the body of the sphinx caterpillar and finally destroys it.

No. 136.—The White-lined Deilephila.

Deilephila lineata (Fabr.).

This handsome moth (see Fig. 264) is a comparatively common insect, and has a wide geographical range, being found throughout the greater portion of the United States and Canada, also in the West Indies and in Mexico. It is double-brooded, appearing on the wing early in June, and again in September. Its period of activity begins with the twilight, when it may be seen flitting about with great rapidity, hovering like a humming-bird over flowers while extracting their nectar. The ground color of the fore wings is a rich greenish olive, with a pale-buff stripe or bar extending along the middle of the wing from the base to near the tip; along the outer margin there is another band or stripe nearly equal in width and of a dull-gray color, and the veins are distinctly margined with white. The hind wings are small, and are crossed by a wide, rosy band, which covers a large portion of their surface, while above and below this band the color is almost black, the hinder margin being fringed with white. On the body there is a line of white on each side, extending

FIG. 264.

from the head to the base of the thorax, where it unites with another line of the same color, which extends down the middle, and, dividing, sends a branch to each side. The abdomen is

FIG. 265.

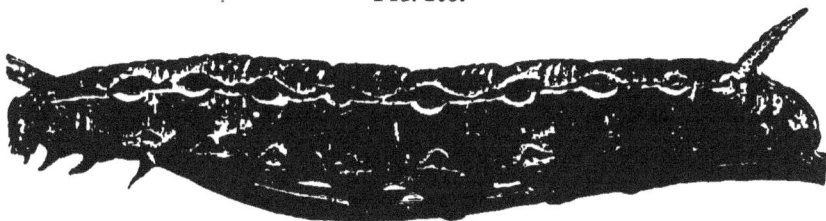

greenish olive spotted with white and black; the wings, when expanded, measure about three and a half inches across.

The larva is found occasionally feeding on the leaves of the grape-vine, but more commonly on purslane; it feeds also on turnip, buckwheat, and apple leaves. It is very variable in color. The most common form is that shown in Fig. 265, where the body is yellowish green, with a row of prominent spots along each side, each spot consisting of two curved black lines, enclosing a crimson patch above and a pale-yellow line below, the whole being connected by a pale-yellow stripe edged with black. In some instances these spots are disconnected, and the space between the black crescents is of a uniform cream-color. The breathing-pores, lower down the side, are margined with black, or black edged with yellow. The other form of the caterpillar is black, with a yellow line down

FIG. 266.

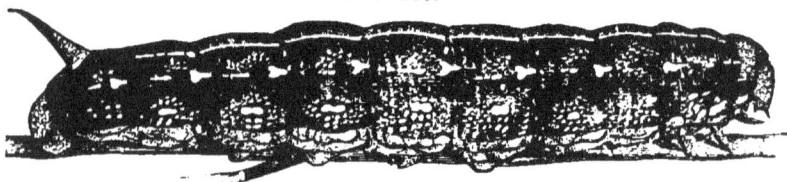

the back, and a double series of yellow spots and dots along the sides. It is shown in Fig. 266.

When mature, it buries itself under the surface, where, within a smooth cavity, it changes to a light-brown chrysalis, the moth emerging early in September, when it deposits eggs, from which the second brood of larvæ are produced, which mature, enter the ground, and change to chrysalids before winter sets in.

Since it feeds mainly on plants of little value, and on these in no great abundance, it is scarcely entitled to be classed with injurious insects; yet on account of its being found occasionally feeding on grape leaves it is deserving of mention here. A two-winged parasitic fly, a species of Tachina, infests it and destroys a large number of the larvæ.

No. 137.—The Dark-veined Deilephila.

Deilephila chamænerii Harris.

This moth very closely resembles the white-lined Deilephila, No. 136, as will be seen from Fig. 267. It has the same greenish-olive color, and almost the same stripes and markings; but there are differences which will enable any one with ease to separate the two species. *Lineata* is much the larger insect, measuring, when its wings are spread, about three and a half inches, while *chamænerii* rarely exceeds two inches and three-quarters. The central band on the fore wings in *chamænerii* is wider and more irregular, the thorax also is less marked with white; but the most striking point of difference is that the veins of the fore wings in *lineata* are distinctly lined with white, a characteristic wanting in *chamænerii*.

The mature larva measures from two and a half to three inches in length. The head is small, dull red, with a black

FIG. 267.

stripe across the front at base. The body above is deep olive-green, with a polished surface; there is a pale-yellowish line along the back, terminating at the base of the caudal horn, and on each segment, from the third to the twelfth inclusive, there is a pale-yellow spot on each side, about half-way between the dorsal line and the breathing-pores, largest on the segments from the sixth to the eleventh inclusive; the spot on the twelfth segment is elongated, and, extending upwards, terminates at the base of the horn. There is a wide but indistinct blackish band across the anterior part of each segment, in which the yellow spots are placed, and the sides of the body below the spots are thickly sprinkled with minute raised yellow dots. The horn is long, curved back-wards, red, tipped with black, and roughened on its surface;

17

the breathing-pores oval, yellow, and margined with dull black.   Under surface pale pinkish green, feet black, prolegs pink, with a patch of black on the outside of each.

This description of the larva was taken from three specimens found feeding on a grape-vine early in July.   One of them matured and formed a slight cocoon of leaves fastened with silken threads on the surface of the ground, after the manner of the green grape-vine sphinx, No. 132; the other two died before completing their transformations.   This larva is said to feed also on purslane; it is not nearly so common as *lineata*, and is not likely ever to prove injurious to any considerable extent.

No. 138.—The Beautiful Wood-nymph.
Eudryas grata (Fabr.).

The larva of this lovely moth is quite destructive to the foliage of the vine, upon which the moth itself is often found resting during the daytime, its closed wings forming a steep roof over its back, and its fore legs, which have a curious muff-like tuft of white hairs, protruded, giving the insect a very singular appearance.   When its wings are expanded, they measure about an inch and three-quarters across.   (See Fig. 268.)   Its fore wings are creamy white, with a glossy

Fig. 268.

surface; a wide brownish-purple stripe extends along the anterior margin, reaching from the base to a little beyond the middle of the wing, and on the outer margin is a broad band of the same hue, widening posteriorly, and having a wavy white line running through it, formed by minute pearly dots or scales, and a dull deep-green edging on its inner side. The brownish-purple band is continued along the hinder edge, but gradually becomes narrower, and terminates when near the base. There are also two brownish spots near the middle of the wing, one round, the other kidney-shaped; these are sometimes so covered with pearly-white scales as to be indistinct above, but are clear and striking on the under side. The hind wings are deep yellow, with a broad brownish-purple band along the hinder margin, extending nearly to the outer angle, and powdered with a few pearly-white scales; there is a faint dot on the middle of the wing, which is more prominent on the under side. The head is black, and there is a wide black stripe down the back, merging into a series of black spots extending to near the tip of the abdomen, which is tufted with white. The shoulder-covers are white, and the sides of the body deep yellow, with a row of black dots along each side close to the under surface. The wings beneath are reddish yellow, and the body white. The moth appears during the latter part of June or early in July, and is active at night.

The eggs are laid on the under side of the leaves, singly or in small groups, and are among the prettiest of insect eggs; they are circular and very flat (see *e* and *f*, Fig. 269), about one-thirtieth of an inch in diameter, and less than half of that in thickness. They are yellowish, or greenish yellow, and are beautifully sculptured with radiating ribs from a central round dot, the ribs interlaced with gracefully curving lines.

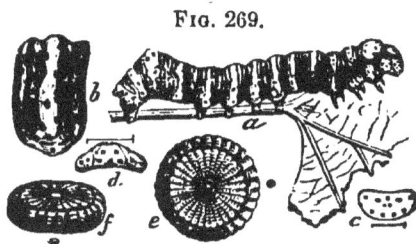

Fig. 269.

On escaping from the egg, the young larvæ are yellowish green, dotted with black; they eat small holes in the leaves, and, when at rest, throw the hinder segments of the body forward over the anterior ones, making a curious sort of loop; as they grow larger they devour all parts of the leaf, the framework as well as the softer substance. When mature, they are about an inch and a half long, and appear as shown at *a* in Fig. 269. The body tapers towards the head, and becomes thicker as it approaches the posterior extremity; the head is orange, dotted with black, the body pale bluish, crossed by bands of orange and many lines of black. Each segment, except the head and the terminal one, is crossed by an orange band of nearly uniform width, except that on the twelfth segment, which is wider; on the terminal segment there are two bands. All these bands are dotted more or less with black, a single short brown hair arising from each dot. The number of black lines crossing each segment is usually six; b shows one of the segments magnified; at c the horny shield behind the head is shown; and at d the hump towards the hinder extremity, all enlarged. The breathing-pores are oval and black. The under side is very similar to the upper. Although partial to the vine, it feeds also on the Virginia creeper, and occasionally on the hop.

When full grown, which is usually some time during the month of August or early in September, the larva descends from the vine and seeks some suitable location in which to pass the chrysalis state. It frequently bores into decaying wood, and is fond of taking refuge in corn-cobs; it is also said to burrow under ground sometimes. In confinement it bores readily into pieces of cork, excavating with its jaws a chamber but little larger than the chrysalis which is to rest in it, and when finished the chamber is provided with a cap or cover composed of minute fragments of cork united by a glutinous secretion. On lifting this lid, there will be seen a dark-brown chrysalis, about seven-tenths of an inch long. Sometimes the moth escapes from the chrysalis late in the same season, but commonly it remains in this condition until the following spring.

This insect is subject to the attacks of a two-winged parasite, a species of Tachina, not unlike the common house-fly in appearance. (See Fig. 270, which shows the insect in its three stages of larva, pupa, and fly; also the anterior segments of a caterpillar, with eggs in position.) This parasite is also found on the army-worm and several other caterpillars. It is about a quarter of an inch long, with a white face, large, reddish eyes, a dark, hairy body, four dark lines down the thorax, and patches of a grayish shade along the sides of the abdomen. The parent fly deposits her eggs on the back of the caterpillar, usually a short distance behind the head, securely fastened by a glutinous substance secreted with them. From

Fig. 270.

these hatch tiny grubs, which eat their way into the body of the caterpillar, feed upon its substance, and finally destroy it, the grubs, when mature, escaping from the body of their victim and changing to oval, smooth, dark-brown pupæ. Usually a large proportion of the caterpillars are infested by this friendly parasite; otherwise they would soon become a source of much annoyance to grape-growers.

Where artificial remedies are required, the vines may be syringed with hellebore and water or Paris-green and water, as directed for the larva of No. 140. Hand-picking may also be resorted to.

No. 139.—The Pearl Wood-nymph.

Eudryas unio (Hübner).

This is a very near relative of *Eudryas grata*, No. 138, and so closely do the two species resemble each other in the larval condition that it is difficult to distinguish between them. *Unio* has usually been regarded as a grape-feeding insect, but from recent observations of Mr. Lintner, of Albany, New York, who has found and reared the larva on an entirely different plant, *Euphorbia coloratum*, it is possible that it may not feed on the grape-vine at all, and that Dr. Fitch, who first announced this as its food-plant, may have mistaken the larva of *E. grata* for *unio*. Since there seems to be some doubt about the matter, we shall briefly describe the insect here.

The moth (Fig. 271) is a little smaller than *grata,* measuring, when expanded, about one inch and three-eighths.

FIG. 271.

It differs also in the following par-
ticulars : on the fore wings the
brownish-purple stripe on the front
margin is extended farther along
the wing, the bordering of the outer
margin is paler and more uniform in
width, the inner edge is wavy instead
of straight, and the bordering of the hind margin is wider and
more distinct. The border on the hind wings is much paler,
and extends the whole length of the outer margin.

The larva is nearly an inch and a quarter long. The head
is of an orange color, spotted with black, the body banded
with white, black, and orange, most of the segments having
three white and three black lines on each side of a central
orange band. The body tapers towards the head, the hinder
segments being elevated.

The chrysalis is reddish brown, with rows of very minute
teeth on the back, and a thick, blunt spine on each side of
the abdomen at the tip.

No. 140.—The Eight-spotted Forester.

Alypia octomaculata (Fabr.).

While the moth of this species is very different in appearance from Nos. 138 and 139, the larva is very similar, being
white or pale bluish, with many black lines, and an orange
band across each segment. This larva (Fig. 272, *a*) may,
however, be distinguished by its having *eight* black lines on
each segment (counting the two which border the orange band)
(see *b*, Fig. 272) instead of *six;* it has also a series of white
spots along each side close to the under surface. The orange
bands are fainter on the anterior segments, and those on the
middle segments are dotted with black, and from each of
these dots there arises a short whitish hair. The head and
the upper part of the next segment are of a deep orange,

with black dots and a polished surface. When young, the larva is paler, with less distinct markings; it feeds on the under side of the leaf, and when alarmed can let itself down to the ground by a silken thread, regaining its position by the same thread when the danger is past. When nearly full grown, it sometimes conceals itself during the daytime within a folded leaf.

FIG. 272.

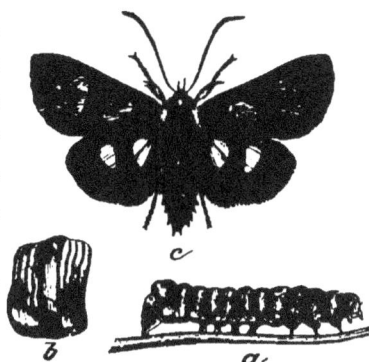

Before effecting its next change, it moulds for itself an earthen cell, upon or just below the surface, which is not lined with silk, and within this enclosure is transformed into a brown chrysalis, from which, in the early brood, the moth escapes in a few days. There are usually two broods each year, the moths appearing on the wing in May and August, the caterpillars in June and July and in September.

The moth is shown at *c* in the figure. It is a very beautiful creature, of a deep blue-black color, with two large pale-yellow spots on each of the front wings, and two white spots on each of the hind wings. In the figure the female moth is represented; the male has the spots on the wings proportionately larger, and a conspicuous white mark along the tip of the abdomen. The shoulder-covers are yellow, and the legs partly orange. The wings, when spread, measure from an inch to an inch and a quarter or more across.

This insect is very generally distributed, being found in most portions of the United States and Canada. Where the larva proves destructive, it may be subdued by syringing the foliage with Paris-green and water, in the proportion of a teaspoonful to two gallons, or powdered hellebore and water, in the proportion of one ounce to two gallons.

No. 141.—The Grape-vine Epimenis.

Psycomorpha epimenis (Drury).

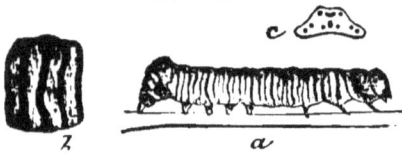

There is still another grape-feeding insect which, in the caterpillar state, bears a strong general resemblance to Nos. 138 and 139. The larva (Fig. 273, *a*) in this species is smaller, of a bluish-white color, with four transverse black bands on each segment, as shown at *b* in the figure, and a few black dots, but lacks the orange bands which distinguish the three species last described. The shield behind the head, the hump on the twelfth segment, and the anal plate are of a dull-orange color; the dots on the hump are arranged as shown at *c* in the figure. The young larva attacks the terminal buds of the vine in spring, fastening the young leaves by a few silken threads, and secreting itself within the enclosure. When full grown, which is usually towards the end of May, it bores into soft wood or any other suitable substance, and there changes to a reddish-brown chrysalis, about four-tenths of an inch long, roughened on the joints, and having a curious, flattened, horny projection on each side of the tip. Within this enclosure it remains until the following spring, when the perfect insect escapes.

FIG. 273.

The moth (Fig. 274) is of a velvety-black color, with a broad, irregular, white patch extending nearly across the front wings, and a somewhat larger and more regularly formed spot of a dull orange-red across the hind wings. The wings are also sprinkled with brilliant purplish scales, most numerous along the outer margins, where they form a narrow band. The under side is paler, with similar markings, the purplish scales appearing very distinct on the front and posterior margins of the hinder wings. The antennæ of the male are toothed,

FIG. 274.

those of the female thread-like. Fig. 274 represents the male. Should this insect ever prove destructive, it may be subdued by the treatment recommended for No. 140, the species last described.

No. 142.—The American Procris.

Procris Americana Harris.

The larvæ of this destructive insect feed in flocks, arranged in a single row on the under side of the vine leaves, as shown in Fig. 275. The egg-clusters from which these larvæ proceed, consisting of twenty eggs or more, are fastened by the moth to the under side of the leaves. While young, the little caterpillars eat only the soft tissues of the leaves, leaving the fine net-work of veins untouched, as shown on the right of the figure, but as they grow older they devour all but the larger veins, as shown on the opposite

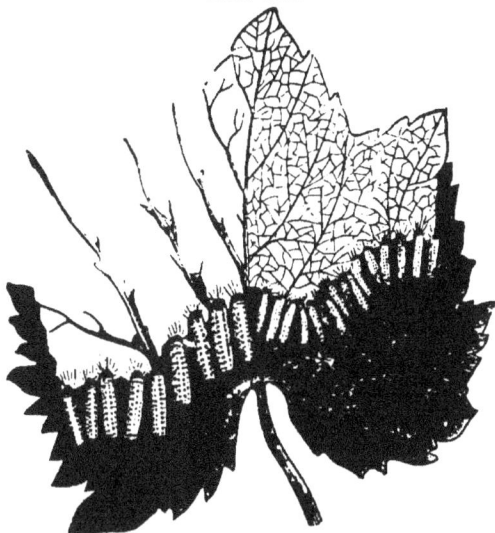

FIG. 275.

side. They acquire full growth in August, when they measure about six-tenths of an inch in length, are of a yellow color, slightly hairy (see Fig. 276, *a*), with a transverse row of black spots on each segment; they feed with their heads towards the margin, and gradually retreat as the leaf is devoured. When full grown, they disperse, and, retiring to some sheltered spot or crevice, construct their tough, oblong-oval cocoons, one of which is shown at *c* in the figure, within which in about three days they change to shining brown chrysalids (*b*) about three-tenths of an inch long, from which the moths escape in about ten or twelve days, and soon deposit eggs for the second brood, which mature later in the season.

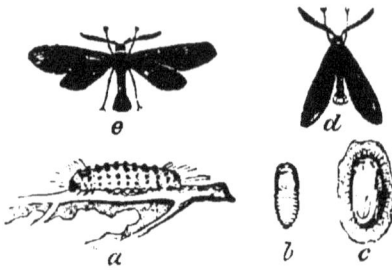

Some few of them produce moths before winter approaches, but the greater portion remain in the chrysalis condition during the winter, the moths escaping the following June.

FIG. 276.

The moth is of a blue-black color, with an orange-yellow collar, and a notched tuft at the extremity of the body; the wings are very narrow, and when expanded measure nearly an inch across. In Fig. 276, *e* represents the moth with the wings spread, *d* the same with the wings closed. This insect is more common in the West and South than in the East, and is sometimes very injurious.

They may be destroyed by syringing the vines with Paris-green and water, as recommended for No. 140. There is a small parasite, a black, four-winged fly, which attacks this larva and destroys it.

No. 143.—The Grape-vine Leaf-roller.

Desmia maculalis Westwood.

This insect, although most abundant in the Southern States, is very generally distributed, and will, no doubt, in its caterpillar form be familiar to most grape-growers. In Fig. 277, 1 represents the larva, natural size, 2 a magnified view of a portion of the anterior part of its body, 3 the chrysalis, 4 the male moth, 5 the female moth.

The moth is a very pretty little creature, measuring, when its wings are expanded, about nine-tenths of an inch or more across. The wings are dark brown, nearly black, with a coppery lustre, and lightly fringed with white; the fore wings have two white spots, nearly oval in form, the hind wings but one white spot in the male, which is usually divided, forming two, in the female. The body is black, crossed in the female by two white bands, in the male by one only. The male moth has the antennæ elbowed and thickened near the middle, in the female they are uniform and thread-like.

FIG. 277.

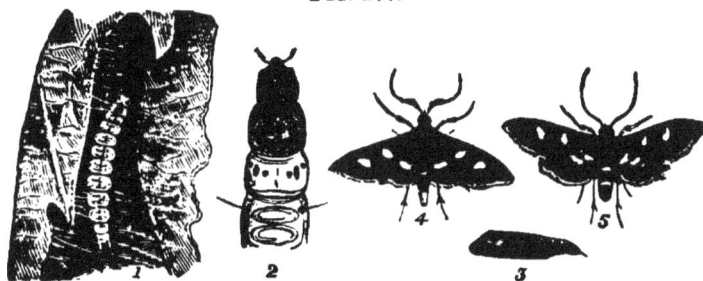

There are two broods of the insect during the summer. The first moths, which have passed the winter in the chrysalis state, appear early in June, and deposit their eggs singly on the leaves of the vine, which are soon hatched, the young worm at once manifesting its leaf-folding propensities by turning down a small portion of the leaf on which it is placed and living within the tube thus formed. As it increases in size, a larger case is made, often the whole leaf being rolled into a large cylinder, wider at one end than at the other, and firmly fastened with stout silken threads. In this hiding-place the little active wriggling creature lives in comparative safety, issuing from it to feed on the surrounding foliage. It is so very rapid in its movements, both backwards and forwards, that it frequently escapes detection by suddenly slipping out of its case when disturbed and falling to the ground. The length of the full-grown caterpillar is about three-quarters of an inch; the body is yellowish green at the sides, a little darker above, glossy and semi-transparent, with a few fine yellow hairs on each segment. The head is reddish yellow, and the next segment behind it has a crescent-shaped patch above of the same color; on the third segment there are two or three black spots on each side, and on the twelfth segment one. The first brood of caterpillars are full grown about the last of July, when they change to chrysalids, from which the moths escape early in August; the second brood of larvæ are found on the vines in September.

The chrysalis (3, Fig. 277) is about half an inch long and of a dark-brown color. It is usually formed within the folded leaf; hence the last brood which pass the winter in this inactive state may, in a great measure, be destroyed by carefully going over the vineyard late in the season, before the leaves fall, and picking off the folded leaves and burning them; or the larvæ may be destroyed earlier in the season by crushing the folded leaves, taking care that the active occupants do not escape. Although this insect is usually common, it is seldom very destructive anywhere.

No. 144.—The Gartered Plume-moth.

Oxyptilus periscelidactylus (Fitch).

The family of moths to which this insect belongs are called plume-moths, from their having the wings divided into feather-like lobes.

The larva (Fig. 278, *a*) appears on the grape-vines in spring, as soon as the young foliage has fairly started, fastening the terminal leaves into a spherical form, and living within the enclosure, where it feeds on the tender leaves and young bunches of blossom. It is usually solitary in its habits, but sometimes two or three are found together. When full grown, which is usually early in June, it is about half an inch long, and is of a yellowish-green color, with transverse rows of dull-yellow tubercles, from each of which arises a small tuft of white hairs. There is a line down the back of a deeper green, and the body is paler between the segments. The head is small, yellowish green, with a band of black across the front; feet black, tipped with pale green; the pro-legs, which are long and thin, are greenish. When matured, it spins a few silken threads on the under side of a leaf, or in some other convenient spot, and, having entangled its hind legs firmly in the web of silk, sheds its hairy skin and becomes a chrysalis.

An odd-looking little thing it is (see Fig. 278, *b*), about four-tenths of an inch long, angular and rugged, and when touched it wriggles about very briskly. It has two rather long, compressed horns placed side by side, extending upwards, on the middle of its back; one of these is shown, enlarged, at *c;* it has also other smaller projecting points and ridges. At first its color is pale yellowish green, but it soon grows darker, becoming reddish brown, with darker spots. It remains in this condition from one to two weeks, when the perfect insect appears.

FIG. 278.

The moth, which is shown in the figure at *d*, is an elegant little insect, its wings measuring, when expanded, about seven-tenths of an inch across. The fore wings are long and narrow, and cleft down the middle about half-way to their base, the posterior half of the wing having a notch in the outer margin. Their color is yellowish brown, with a metallic lustre, and several dull-whitish streaks and spots. The hind wings are similar in color to the anterior pair, and are divided into three lobes; the lower division is complete, extending to the base, the upper one not more than two-thirds of the distance. The outer and hind margins of the wings, as well as all the edges of their lobes, are bordered with a deep whitish fringe, sprinkled here and there with brown; the body is long and slender, and a little darker than the wings. The antennæ are moderately long and thread-like, nearly black, but beautifully dotted with white throughout their whole length. The legs are long, banded alternately with yellowish brown and white, the hind ones ornamented with two pairs of diverging spines, having at their base a garter-like tuft of long brown scales, from which feature the moth derives its name.

This insect is single-brooded; it is common throughout Ontario and Quebec. Where troublesome, it may be subdued by hand-picking, or by pinching the clusters of leaves and crushing the larvæ.

No. 145.—The Grape-vine Cidaria.

Cidaria diversilineata Hübn.

This is a pretty yellow moth, producing a geometric or looping caterpillar which consumes the foliage of the vine. The insect passes the winter in the caterpillar state, hibernating in some secure retreat until aroused to activity by the warmth of spring, when, after feeding a few days on the young vine leaves, it becomes a chrysalis, producing the moth about ten days afterwards. The moths within a few days deposit their eggs on the leaves of the vine, which hatch early in June, and the larvæ nearly complete their growth by the end of the month, pass into the chrysalis state, and appear as moths again in July and August. These latter deposit eggs for the second brood of larvæ, which, before reaching maturity, become torpid, and remain in this condition until spring.

The moth (Fig. 279) measures, when its wings are expanded, about an inch and a half across. Its color is pale ochre-yellow, crossed by many grayish-brown lines, and clouded

Fig. 279.

with patches of the same, particularly along the margin of the wings. The body and legs are similar in color to the wings, the latter being marked with black about the joints.

Early in June the reddish geometric caterpillars of this moth are found upon the leaves, out of which they eat numerous pieces of various sizes and shapes. By the middle of the month they become full

grown, when they measure about an inch and a quarter
long. (See Fig. 280.) The head is dull reddish brown, the
body yellowish green, with a few
small whitish dots on each segment.

FIG. 280.

On each side of the second segment
is a small reddish spot, and on the
third a larger one of a darker shade;
on this latter segment there is a fold in the skin, which makes
the spot appear as a brown prominence. The terminal segment is furnished with two short, greenish spines, which
extend backwards; the surface of the body is wrinkled; the
under surface reddish, with a central reddish line, bordered
with white, which is margined with dull red. These larvæ
are very variable in color, being sometimes yellowish green,
whitish green, deep red, and occasionally dark brown, nearly
black. When alarmed, they straighten themselves out, and
remain for some time without moving, when, being so nearly
of the color of the twigs they rest on, they usually escape
detection.

Where these larvæ are sufficiently numerous to prove
troublesome, the vines may be syringed with Paris-green and
water, or hellebore and water, as recommended for No. 140.

No. 146.—The Yellow Woolly-bear.

Spilosoma Virginica (Fabr.).

This common caterpillar is so well known that it is scarcely
necessary to describe it. Every one who has a garden in
which fruits or flowers are grown must have frequently met
with it, for no insect is so uniformly common and troublesome
as this one. It seems to have a special liking for the leaves
of the grape-vine, but it feeds also on the leaves of a great
variety of plants, shrubs, and trees.

The moth from which the larva is produced is shown at *c*,
Fig. 281, and is commonly known as the "white miller."
It passes the winter in the chrysalis state, and appears on the
wing late in April or early in May, and, when its wings are

expanded, measures from one and a half to two inches across. The figure represents a female; the males are somewhat smaller. Both sexes have the wings white, with a few black dots, which vary in number in different specimens; in some there are two on each of the front wings, and three on each of the hinder pair; in others the spots are partly or almost entirely wanting. The dot, however, near the middle of the front wings is almost always present, although sometimes very faint. The under side usually has the spots more dis-

Fig. 281.

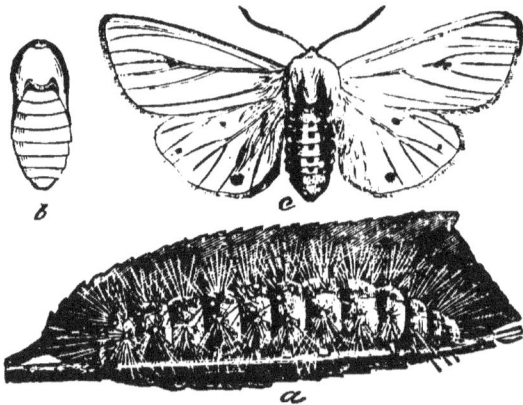

tinct than the upper, and sometimes there is a slight tinge of yellow over its white surface. The antennæ are white above, dark brown below, the head and thorax white, and the abdomen of an orange color, usually streaked across with white, and having three rows of black spots, one above and one on each side. The under side of the abdomen is white, occasionally tinged with orange, and the thighs of the fore legs ochre-yellow.

The eggs, which are round and yellow, are deposited on the under side of the leaves in large clusters, and in a few days hatch into small hairy caterpillars, which feed for a time in company, devouring at this tender age the under side of the leaf only, the outer skin over the eaten part soon becoming yellow and withered. When partly grown, they separate, each one choosing his own course, and by this time their digestive powers have become sufficiently strong to enable them to eat freely of all parts of the leaf.

The full-grown caterpillar (Fig. 281, *a*) is nearly two inches long, and usually of a yellowish color, but the color varies greatly, and in the same brood there may be found with the yellow some straw-colored and others brown, from a light to a very dark shade. On each segment there are a number of yellowish tubercles, from each of which there arises a tuft of hairs of a yellowish or brownish color, sometimes intermingled with a few black ones. The spaces between the segments are crossed by dark-brownish or sometimes black lines, and there is a line of the same color along each side; the under surface of the body is dark also. When mature, it seeks some sheltered nook or cranny in which to pass the chrysalis state, and, having found a suitable location, proceeds to divest its body of the hairy covering, and with this woven together by silken threads it constructs a slight cocoon, within which the chrysalis is formed, of a chestnut-brown color, as shown at *b* in the figure. There are at least two broods of this insect each year, and these broods so intermingle that the insect may almost always be found in one or other of its stages from May to October.

Fig. 282.

This species is subject to the attack of several kinds of Ichneumon flies, which destroy immense numbers of them every year; one of these, *Ophion bilineatus* Say, is represented in Fig. 282. Were it not for these friendly agencies constantly at work the common woolly-bears would soon become very destructive. As it is, they are sometimes very injurious; when this is the case, hand-picking

should be resorted to, and if this is done while the larvæ are
young and feeding in company, their destruction is easily
accomplished.

No. 147.—The Pyramidal Grape-vine Caterpillar.

Pyrophila pyramidoides (Guen.).

This caterpillar (Fig. 283) is frequently destructive to
grape-vines, particularly to those grown under glass, and may
be found on the leaves full grown about the middle of June.
It is nearly an inch

FIG. 283.

and a half long, the
body tapering to-
wards the front, and
thickened behind.
The head is rather
small, of a whitish-
green color, with the mandibles tipped with black; the body
whitish green, a little darker on the sides, with a white stripe
down the back, a little broken between the segments or rings,
and widening behind. There is a bright-yellow stripe on
each side close to the under surface, which is most distinct on
the hinder segments, and a second one of the same color, but
fainter, half-way between this and the dorsal line; this latter
is more distinct on the posterior portion of the body, and
follows the peculiar prominence on the twelfth segment, as
shown in the figure. The under side of the body is pale
green.

When full grown, the caterpillar descends to the ground,
and, drawing together some loose fallen leaves or other
rubbish, spins a slight cocoon, within which it changes to a
dark-brown chrysalis, from which the perfect insect escapes
in the latter part of July.

The moth (Fig. 284) measures, when its wings are expanded,
about one and three-quarter inches. The fore wings are dark
brown shaded with paler brown and with dots and wavy lines
of dull white; the hind wings are reddish, with a coppery

lustre, becoming brown on the outer angle of the front edge of the wing, and paler towards the hinder and inner angle.

The under surface of the wings is much paler than the upper. The body is dark brown, its hinder portion banded with lines of a paler hue.

Fig. 284.

While partial to the grape, the larva feeds also on thorn, plum, raspberry, red-bud, *Cercis Canadensis*, poplar, and probably other trees, shrubs, etc. The insect is distributed over a wide area. Where they are numerous enough to prove troublesome, they may be collected and destroyed by jarring the trees or vines on which they are feeding, when they will drop to the ground.

No. 148.—The Silky Pyrophila.

Pyrophila tragopoginis (Linn.).

The caterpillar of this moth is of a yellowish-green color, with a few very fine brownish hairs scattered over the upper surface of its body. It is found feeding on the grape-vine, and sometimes in sufficient numbers to become a source of annoyance; it attains full growth about the middle of June, when it measures an inch and a quarter or more in length. The head is small, green, the jaws tipped with brown; the upper surface of the body is yellowish green, a little paler between the joints; there is a white stripe down the back, and two of the same color along each side, the lowest one being most distinct. On each segment there are several small whitish dots, from each of which arises a single fine hair. The under side is deeper in color than the upper. When mature, it changes to a brown chrysalis, a little under the surface of the ground, from which the perfect insect escapes in July.

The moth measures, when its wings are spread, about an

inch and a quarter across.   Its fore wings are grayish brown
with a silky lustre, with several pale dots on the front edge,
and three short dark streaks near the middle.   The hind
wings are paler.

When found to be injurious, the caterpillars may be subdued
by hand-picking.

No. 149.—The Spotted Pelidnota.

Pelidnota punctata (Linn.).

This enemy to the grape-vine is a large and handsome
beetle (Fig. 285, *c*), which eats the leaves, making numerous

Fig. 285.

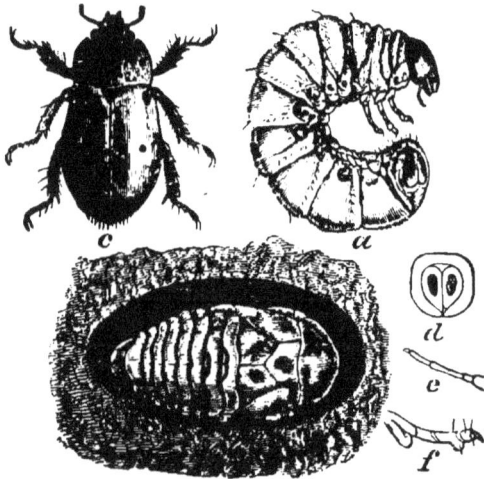

holes in them.   It measures about an inch in length and half
an inch in width at its widest part, is nearly oval in form, of
a dull reddish-yellow color, with a polished surface, and three
black spots on the outer side of each wing-cover.   The thorax, which is rather darker than the wing-covers, is slightly
bronzed, and has a small black dot on each side; the jaws and
hinder part of the head are black, so also is the scutellum, a
small, nearly triangular piece at the point of juncture of the
wing-covers with the thorax.   The transparent, gauzy wings,

which are concealed under the wing-cases when not in use, are dark brown. The under side of the beetle is dark green, with a metallic lustre, downy about the middle, with fine brownish hairs. Legs, dark shining green. It appears during July and August, and is active during the day, flying from vine to vine with a heavy, awkward flight and a loud, buzzing noise. The female deposits her eggs in rotten wood, on which the larva, when hatched, feeds ; the decaying stumps and exposed decaying roots of pear, hickory, and other trees being selected for this purpose.

When full grown, the larva measures nearly two inches in length, and presents the appearance shown at *a* in the figure. It has a chestnut-brown head and a translucent, white body, and much resembles the larva of the May-beetle, No. 113, but is of a clearer white color, and has a heart-shaped swelling on the terminal segment, which is short and cut off squarely. A front view of the markings on this segment is given at *d* in the figure. When mature, it forms a slight cocoon, into which are woven its own castings mixed with particles of the surrounding wood, and within this it changes to a pupa, as seen at *b*, from which the beetle escapes about ten days afterwards ; *e* represents the antenna of the larva, and *f* one of its legs, both magnified.

This insect is common throughout the Eastern and Western States and the central portions of Canada. Should it at any time prove injurious, it can easily be reduced in numbers by hand-picking. It feeds also on the Virginia creeper, *Ampelopsis quinquefolia.*

No. 150.—The Grape-vine Flea-beetle.

Graptodera chalybea (Illig.).

This pretty but destructive little beetle (see Fig. 286) forces itself upon the attention of grape-growers very prominently in the spring season, when, awakened by the reviving warmth of the sun from its winter state of torpidity, and with appetite sharpened by its long fast, it commences its work of destruction by eating away the substance of the buds as soon as they begin to swell, thus destroying many bunches of

Fig. 286.

Fig. 287.

grapes in embryo. It goes on with this work for about a month, when it gradually disappears. Before leaving, however, the beetle provides for the continuance of its race by depositing little clusters of orange-colored eggs on the under side of the young vine leaves, which in a few days produce colonies of small, dark-brown larvæ, which feed on the upper side of the leaves, riddling them, and when numerous they devour the whole leaf except the larger veins, and sometimes entirely strip the vines of foliage. Fig. 287 represents the larvæ in various stages of growth at work on the vine, accompanied also by some of the beetles.

In three or four weeks the larva attains full growth, when it is a little more than three-tenths of an inch long, usually of a light-brown color, sometimes dark, and occasionally paler and yellowish. The head is black, and there are six or eight shining black dots on each of the other segments of the body, each dot emitting a single brownish hair. The under surface is paler than the upper, its feet, six in number, are black, and there is a fleshy, orange-colored proleg on the terminal segment. It is shown magnified in Fig. 288.

When mature, the larvæ leave the vines and descend to the ground, where they burrow under the earth and form small, smooth, oval cells, within which they change to dark-yellowish pupæ. After remaining two or three weeks in this condition, the beetles issue from them, and the work of destruction goes on; but since they live at this season of the year altogether on leaves, of which there is an abundance, the injury done is much less than in the spring.

Fig. 288.

The beetle is about three-twentieths of an inch long, and varies in color from a polished steel-blue to green, and occasionally to a purplish hue, with a transverse depression across the hinder part of the thorax. The under side is dark green, the antennæ and feet brownish black; the thighs are stout and robust, by means of which the insect is able to jump about very nimbly. One of the legs, detached from the body, is shown in Fig. 286. On the approach of winter the beetles retire to some suitable shelter, as under leaves, pieces of bark, or in the earth immediately around the roots of the vines, where they remain inactive until the following spring. Besides the vine, they feed on the Virginia creeper, *Ampelopsis quinquefolia,* and the alder, *Alnus serrulata,* and sometimes eat the leaves of the plum-tree.

Remedies.—To destroy the beetles it is recommended to strew in the autumn air-slaked lime or unleached ashes around the infested vines, removing and destroying all rubbish which might afford shelter. In the spring the canes and young foliage may be syringed with water in which has been stirred a teaspoonful of Paris-green to each gallon. Strong soap-suds have also been recommended, and are deserving of trial. On chilly mornings the beetles are comparatively sluggish and inactive, and may then be jarred from the vines on sheets and collected and destroyed. These insects are much more abundant in some seasons than in others.

No. 151.—The Rose Beetle.

Macrodactylus subspinosus (Fabr.).

This beetle, commonly known as the rose-bug, attacks the rose, and is also very injurious to the grape-vine, the apple, cherry, peach, plum, etc. Its body (see Fig. 289) is a little more than one-third of an inch long, slender, and tapering a little towards each extremity. Its color is dull yellowish when fresh, arising from its being covered with a grayish-yellow down or bloom, and its long, sprawling legs are of a dull pale-reddish hue, with the joints of the feet tipped with black and armed with very long claws. The down on the body of the beetle is easily rubbed off, producing quite a change in its appearance, the head, thorax, and the under side of its body becoming of a shining black.

FIG. 289.

These beetles sometimes appear in swarms about the time of the blossoming of the rose, which in the Northern United States and Canada is usually during the second week in June; they remain about a month, at the end of which period the males become exhausted, drop to the ground, and perish, while the females burrow under the surface, deposit their eggs, then reappear above ground, and shortly afterwards die also.

Each female lays about thirty eggs, which are buried in the earth to the depth of from one to four inches; the eggs are about one-thirtieth of an inch in diameter, whitish, and nearly globular. In about three weeks they hatch, and the young larvæ at once begin to feed on such tender roots as are within their reach. They attain full growth in the autumn, when they are about three-quarters of an inch long and abou.

an eighth of an inch in diameter, of a yellowish-white color, with a tinge of blue towards the hinder extremity, which is thick, obtuse, and rounded ; the head is pale red and horny, and there are a few short hairs scattered over the surface of the body. In October the larva descends below the reach of frost, and passes the winter in a torpid state; in the spring it approaches the surface and forms for itself a little oval cell of earth, within which it is transformed to a pupa during the month of May.

In form the pupa bears some resemblance to the perfect insect, and is of a yellowish-white color, its whole body being enclosed in a thin film that wraps each part separately. In June this filmy skin is rent, when the enclosed beetle withdraws its body and limbs, bursts open its earthen cell, and forces its way to the surface of the ground, thus completing its various stages within the space of one year.

Although these insects have many natural foes, such as carnivorous ground-beetles, insectivorous birds, domestic fowls, toads, etc., they often need the intervening hand of man to keep them within due bounds. When numerous, they may be detached from the vines with a sudden and violent jar, falling on sheets spread below to receive them. They are naturally sluggish, do not fly readily, and are fond of congregating in masses on the foliage they are consuming, and hence in the morning, before the day becomes warm, they can be easily shaken from their resting-places, collected, and burnt, crushed, or thrown into scalding water. This insect is very partial to the Clinton grape, and, where this is to be had, will congregate on it in preference to other varieties, a peculiarity which may be made use of by planting Clinton vines as a decoy, and thus materially lessening the labor involved in the destruction of the beetles.

No. 152.—The Grape-vine Fidia.

Fidia longipes (Mels.).

This enemy to the grape-vine is a chestnut-brown beetle (see Fig. 290), about a quarter of an inch long, with its body densely covered with very short whitish hairs, which give it a hoary appearance. It is first seen in June, and by the end of July has usually disappeared. Its mode of operation is to cut straight, elongated holes about one-eighth of an inch in diameter in the leaves, and when the insects are numerous these are so thickly perforated as to be reduced to mere shreds. This is said to be one of the worst foes the grape-grower has to contend with in Missouri and Kentucky, where at times it literally swarms, and then almost entirely destroys the foliage of large vineyards. It is a native insect, found in the woods feeding on the wild grape, also on the red-bud, *Cercis Canadensis;* of the vines in cultivation it is said to prefer the Concord and Norton's Virginia. Upon the slightest disturbance, or when danger threatens, it has the habit of doubling up its legs and falling to the ground, where for a time it remains motionless, feigning death in the same manner as the plum curculio. Advantage may be taken of this habit, and the insects collected by placing sheets under the vines and jarring them with the hand. The grape-vine Fidia belongs to the great family *Chrysomelidæ*, which includes the grape-vine flea-beetle, the potato-beetle, and many other injurious species. Of the early stages of this insect nothing is yet known.

FIG. 290.

No. 153.—The Grape-vine Colaspis.

Colaspis brunnea Fabr.

This beetle also belongs to the *Chrysomelidæ*, and injures the vine leaves in a manner similar to that of the species last described, riddling them with small round holes, interspersed with larger irregular ones, in a wholesale manner. It is

nearly one-fifth of an inch long (see Fig. 291), of a pale-
yellowish color, with the body densely punctated, and with
elevated lines on the wing-
covers between the rows
of dots. It is found in
most of the Eastern and
Middle States, and de-
vours also the leaves of
the strawberry; it appears
early in July and during August.

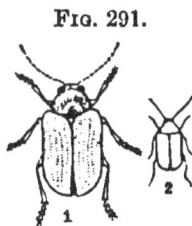

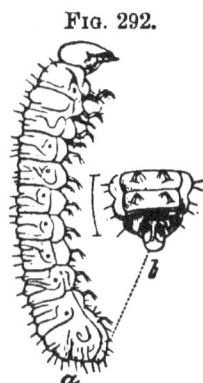

FIG. 291.

FIG. 292.

The eggs are deposited either upon or in
close proximity to strawberry-plants, and
when hatched the young larvæ burrow into
the earth and feed upon the roots of the
strawberry-vines, on which they may be found all through
the fall, winter, and spring months. It is a singular larva,
shown magnified in Fig. 292, and has on the under side
of each of the legless joints a pair of fleshy projections reminding one of legs, each tipped with two or three stiff
hairs. Its body is yellowish or grayish white, with a yellow head. The pupa is formed in the earth during the
month of June, the perfect insect maturing two or three
weeks afterwards.

Remedies.—The beetles may be collected by jarring them
from the vines on sheets early in the morning, and destroyed.
Ashes, soot, or lime applied to the strawberry-vines will in
most instances deter the beetles from depositing their eggs
on them, or will destroy the young larvæ as soon as hatched.

No. 154.—The Red-headed Systena.

Systena frontalis (Fabr.).

This insect belongs also to the *Chrysomelidæ*, and, although
very generally distributed throughout the northern portions
of America, has not until of late been recorded as injurious.
During the summer of 1882, in some parts of the Province
of Ontario, it inflicted much injury on the vines by devour-

ing the green tissues on the upper side of the leaves, causing them to discolor and eventually to wither. This insect is furnished with stout thighs, which enable it to jump like the flea-beetle of the vine, to which it is closely allied. The beetle (Fig. 293) is about one-sixth of an inch in length, the thorax and wing-cases black and densely but very finely punctated. The head is pale red above, between the eyes; the antennæ are rather long and reddish, with the basal joint black. The under side is brownish black. The legs are well adapted for jumping, the thighs being thick and robust.

FIG. 293.

No. 155.—The Light-loving Anomala.

Anomala lucicola (Fabr.).

This insect is a beetle about one-third of an inch long (see Fig. 294), in form resembling the May-beetle, No. 113, which appears late in June or early in July. It is common on both the wild and the cultivated grape-vine, feeding upon the leaves. The beetle is of a pale dull-yellow color, the thorax black, margined with dull yellow, the hind part of the head and the under side of the body also black; sometimes the abdomen is brown.

FIG. 294.

These beetles occasionally appear in swarms, when they devour the foliage very rapidly, the vine leaves soon resembling a piece of net-work, only the large veins, with some of the smaller ones, being left.

Remedies.—Dusting the vines with fresh air-slaked lime, or syringing them with a solution of whale-oil soap or strong tobacco-water, has been recommended. Probably hellebore or Paris-green with water, as recommended for No. 140, would be more effectual.

No. 156.—The Grape-vine Saw-fly.

Selandria vitis Harris.

This is a small four-winged fly (Fig. 295), with a shining black body, except the upper side of the thorax, which is red ; the wings are semi-transparent, and have dark-brown veins, the front pair being clouded, or of a smoky color. The fore legs and under side of the other legs are pale yellow or whitish. The body of the female measures about three-tenths of an inch in length, that of the male somewhat less. The insect is double-brooded, the first brood of flies appearing in the spring, the second late in July or early in August.

FIG. 295.

The eggs are laid on the under side of the terminal leaves of the vine in small clusters, and the larvæ, when hatched, feed in company, side by side, from about half a dozen to fifteen or twenty in a group, preserving their ranks with much regularity, as shown in Fig. 296. They begin at one edge of the leaf and eat the whole of the leaf—including the ribs—to the stalk, and proceed from leaf to leaf down the branch, devouring as they go, until they are full grown. When mature, they measure about five-eighths of an inch in length, are somewhat slender and tapering behind, and thickened before the middle. They are of a pale-yellow color, darker or greenish on the back, with two transverse rows of minute black points across each ring, the head and tip of the last segment being black ; the under side is yellowish. After the last moult the larvæ become entirely yellow, when they leave the vines, descend to the ground, and burrow under its surface. There they form oval cells in the earth, which they line with silk, and within these enclosures change to pupæ, from which the perfect flies escape in about a fortnight. The second brood pass the winter in the pupal state. In Fig. 296 one of the oval

FIG. 296.

cells is shown with the fly resting on it; also one of the pupae.

Occasionally this insect is very destructive, sometimes entirely stripping the vines. In such cases the foliage should be sprinkled with hellebore and water, or Paris-green and water, in the proportions given under No. 140.

No. 157.—The Grape-vine Leaf-hopper.

Erythroneura vitis (Harris).

The accompanying figure, 297, represents the insect commonly known among vine-growers as the "Thrip." The

FIG. 297.

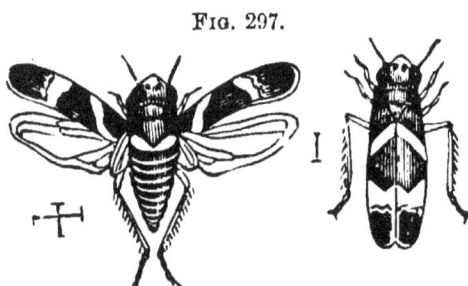

insects are shown magnified; the shorter lines adjoining indicate their natural size. The figure to the left shows the mature insect with its wings expanded, the other the same with its wings closed. It is rather more than one-eighth of an inch long, crossed by two broad, blood-red bands, and a third dusky one at the apex, the anterior band occupying the base of the thorax and the base of the wing-covers, the middle one wide above, narrowing towards the margin. Besides *vitis*, there are half a dozen or more which are supposed to be distinct species, all about the same size, and with the same habits, differing only in the markings on the wings.

These insects pass the winter in the perfect state, hibernating under dead leaves or other rubbish, the survivors becoming active in spring, when they insert their eggs in punctures in the leaves of the vine. The larvæ are hatched during the month of June, and resemble the perfect insect except in size and in being destitute of wings. During their growth they shed their skins, which are nearly white, several times, and, although exceedingly delicate and gossamer-like, the empty skins remain for some time attached to the leaves. The insects feed together on the under side of the leaves, and are very quick in their movements, hopping briskly about by means of their hind legs, which are especially fitted for this purpose. They have a peculiar habit of running sideways, and when they see that they are observed upon one side of a leaf they will often dodge quickly around to the other. They are furnished with a sharp beak or proboscis, with which they puncture the skin of the leaf, and through which they suck up the sap, the exhaustion of the sap producing on the upper surface yellowish or brownish spots. At first these spots are small and do not attract much attention, but as the insects increase in size the discolored spots become larger until the whole leaf is involved, when, changing to a yellow cast, it appears as if scorched, and often drops from the vine. Occasionally the vines become so far defoliated that the fruit fails to ripen.

As the leaf-hopper enters the second stage of its existence, corresponding to the pupal state in other insects, diminutive wings appear, which gradually grow until fully matured, the insect meanwhile becoming increasingly active. With the full growth of the wings it acquires such powers of flight that it readily flies from vine to vine, and thus spreads itself in all directions. It continues its mischievous work until late in the season, when it seeks shelter for the winter.

The Clinton, Delaware, and other thin-leaved varieties suffer more from the attacks of these insects than do the thick, leathery-leaved sorts, such as Concord. These leaf-hoppers are sometimes quite abundant in a vineyard one season and comparatively scarce the next, their preservation depending so much on favorable weather and suitable shelter for the perfect insects during winter.

Remedies.—Various measures have been suggested as remedies. Since the insect does not consume the outer surface of the leaf, it becomes difficult to deal with it. Syringing with strong tobacco-water or soap-suds, or fumigating with tobacco where the vines can be enclosed, so as to prevent the free escape of the smoke, are the most efficient remedies. Dusting with lime, sulphur and lime, hellebore and Cayenne pepper, have all been recommended. Carrying lighted torches through the vineyard at night, the foliage at the same time being disturbed with a stick, will destroy a great many of them, since they fly to the light and are burnt. As a preventive, the ground in the neighborhood of the vines should be kept thoroughly clean, and be several times raked or otherwise disturbed late in the autumn and early in the spring, so as to expose any concealed insects to the killing influence of frost.

A species of bug known as the Glassy-winged Soldier-bug, *Campyloneura vitripennis* Say, feeds on these leaf-hoppers, and devours large numbers of them. Fig. 298 shows this friendly insect in the larval state, and Fig. 299 in the perfect condition. This useful friend, whenever seen, should be protected. In both figures the insect is magnified, the lines at the side showing the natural size. The mature insect is of a pale greenish-yellow color, the head and thorax are tinged with pink, and the upper wings are transparent and ornamented with a rose-colored cross.

Fig. 298.   Fig. 299.

The Grape-leaf Gall-louse.

Phylloxera vitifolia Fitch.

This has been already treated of under the grape phylloxera, No. 125.

Tree-hoppers.

Several insects may be grouped under this name which attack the leaves of the vine, and some of them the succulent branches also.

No. 158.—One of these, the Waved Proconia, *Proconia undata* Fabricius (see Fig. 300), is a cylindrical jumping insect nearly half an inch long, which is said to lay its eggs in single rows in the wood of the canes. Be- Fɪɢ. 300. sides attacking the leaves, this bug punctures with its beak the stems of the bunches of grapes, causing the stems to wither and the bunches to drop off. Sometimes it pumps out the sap so vigorously from the succulent branches that the drops fall in quick succession from its body.

In the southern parts of Illinois this insect is at times very numerous, becoming then one of the worst enemies the grape-grower has to contend with.

No. 159.—The Single-striped Tree-hopper, *Thelia univittata* Harris, is shaped much like a beech-nut, with a perpendicular protuberance on the fore part of its back higher than it is wide, and its summit rounded. The insect is of a chestnut-brown color, tawny white in front, and with a white stripe along the back, extending from the protuberance to the tip. It is about one-third of an inch long and a quarter of an inch in height, and may often be seen on grape-vines in July and August.

No. 160.—Another species is the Black-backed Tree-hopper, *Acutalis dorsalis* (Fitch), a small, triangular, shining insect with a smooth, rounded back. Its color is greenish white, and it has a large black spot on its back, from the anterior corners of which a black line runs off towards each eye; the upper margin of the head and the breast are also black. The female is about one-fifth of an inch long, the male smaller. This species is sometimes found in considerable numbers on grape-vines about the last of July, and a few stragglers usually remain until October.

Tree-bugs.

There are also several species of tree-bugs which infest the vine and suck its juices.

No. 161.—The large green Tree-bug, *Rhaphigaster Pennsylvanicus* De Geer (Fig. 301), is from six to seven tenths of an inch long, flattened in form, of a grass-green color, margined with a light-yellow line, which is interrupted at each joint of the abdomen with a small black spot. The antennæ are black, with some yellow on the basal and terminal joints. It occurs on grape-vines, chiefly in September, and is also found on hickory and willow trees.

Fig. 301.

No. 162.—The Bound Tree-bug, *Pentatoma ligata* Say, is a large green bug closely resembling the species last described, but is more broadly edged all around, except upon its head, with pale red, and has a pale-red spot upon the middle of its back. The antennæ are green, except the three last segments, which are black. This species is a little more than half an inch long, and occurs also on the hazel.

No. 163.—The Modest Tree-bug, *Arma modesta* Dallas, is smaller, being from four to four and a half tenths of an inch long, of a tawny yellowish-gray color, thickly dotted with brown. The wing-cases are commonly red at their tips, and the under glassy wings have a brown spot at their extremities. The under side is whitish, with a row of black dots along the middle of the abdomen, and another on each side. This insect is one of the commonest tree-bugs, and is found in the autumn on a number of different trees and shrubs.

No. 164.—The Grape-vine Aphis.

Siphonophora viticola Thomas.

This species of plant-louse, which is destructive to the leaves and young shoots of the grape-vine, is of a dusky-brown or blackish color, legs greenish, marked with dusky. Most of the lice are wingless, but some have wings clear and glassy, with brownish veins. This is believed to be the same species as that which infests the vine in the southern parts of Europe, viz., *Aphis vitis*, but the insect has not yet been sufficiently studied to decide this with certainty. They cluster in thousands on the ends of the branches, causing the leaves to curl up and the vine to appear very unsightly. They are seen early in the summer, and usually continue but a few weeks, as their enemies, the lady-birds and other predaceous insects, increase so fast as to decimate them within that time. They are common in the South and in the Middle States, but occur only occasionally in the more northern districts.

Should occasion require the application of a remedy, the vines may be syringed with weak lye, tobacco - water, or strong soap-suds.

FIG. 302.

No. 165.—The Broad-winged Katydid.

Cyrtophyllus concavus (Harris).

This is perhaps our commonest species of katydid, and may be distinguished from the other species by the greater breadth and convexity of its wing-covers, which, with their strong midrib and regular venation, much resemble a leaf. The insect (Fig. 302) is about an inch and a half long, its body of a pale green color, with slightly darker wing-cases. The female has a projecting ovipositor or piercer, with which the eggs are thrust into crevices and soft substances. The eggs are of a dark slate color, very flat, pointed at both ends and the edges bevelled : they are about one-eighth of an inch long, and not more than one-third of this in diameter. When in confinement this katydid is said to insert its eggs freely into pieces of cork and other soft substances. The young katydids when hatched, which usually occurs in the following spring, eat almost any tender succulent leaves, and have never been recorded as very injurious. The males are furnished with a pair of musical organs, which they use vigorously as night approaches, and their sharp, shrill notes can be heard at a long distance.

Another and a very similar species is the Oblong-winged Katydid, *Phylloptera oblongifolia* De Geer, which is also said to deposit eggs on grape-twigs.

No. 166.—The Trumpet Grape-gall.

Vitis viticola Osten Sacken.

These are curious, elongated, conical galls, about one-third

Fig. 303.

of an inch long, of a reddish or reddish-crimson color, sometimes inclining to green, growing in considerable numbers on the leaves of the vine. (See Fig. 303.) Though usually found only on the upper surface, they are occasionally seen on the under side also. They are produced by a gall-gnat, an undetermined species of *Cecido-*

myia, and on cutting into the galls they are found to be hollow, each containing a pale-orange larva. It is probable that the larva enters the earth to transform to the pupa, and that the fly is produced the following season.

No. 167.—The Grape-vine Filbert-gall.

Vitis coryloides Walsh & Riley.

In this instance a rounded mass of galls from one and a half to two and a half inches in diameter springs from a common centre at a point where a bud would naturally be found. The mass (see Fig. 304) is composed of from ten to

FIG. 304.

forty opaque, woolly, greenish galls, which have a fleshy, juicy, sub-acid interior, each with a single central, longitudinal cell, one of which is shown at c in the figure, about a quarter of an inch long and one-fourth as wide, containing a solitary orange-yellow larva, about one-eighth of an inch long. This is also the larva of an undetermined species of *Cecidomyia*, a family the members of which may be recognized in the larval state by a peculiar appendage known as a breast-bone attached to the under side near the head. In this species it is almost Y-shaped, as shown at *a* in the figure; the diverging

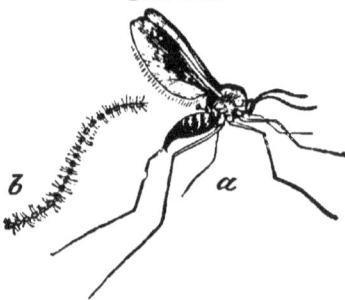

FIG. 305.

branches terminate in two projecting points, which may be extended at will, and which are probably used by the larva in abrading the soft tissues of the gall so as to cause an exudation of sap, on which the larva feeds. The flies belonging to this genus are usually of a dull-black color, like that shown in Fig. 305, *a*, which represents a female fly; the antenna of a male is seen at *b*. The gall is common in July; the larger-sized specimens bear some resemblance to a bunch of filberts or hazel-nuts, hence the name filbert-gall.

No. 168.—The Grape-vine Tomato-gall.

Vitis tomatos Riley.

These galls form a mass of irregular, succulent swellings on the stem and leaf-stalks of the grape-vine (see Fig. 306), very variable in size and shape, from the single, round, cran-berry-like swelling to the irregular, bulbous protuberances which look much like a group of diminutive tomatoes. They have a yellowish-green exterior, with rosy cheeks, and some-times are entirely red; the interior is soft, juicy, and acid. Each gall has several cells, as shown at *a* in the figure, and in each cell there is an orange-yellow larva, which, before the gall has entirely decayed, enters the ground, where it changes to a pupa, and finally emerges as a pale-reddish gnat, with black head and antennæ, and gray wings. This fly also belongs to the family *Cecidomyia*, and is known to entomologists as *Lasioptera vitis* of Osten Sacken.

The larvæ are liable to be attacked by a parasite, and also

Fig. 306.

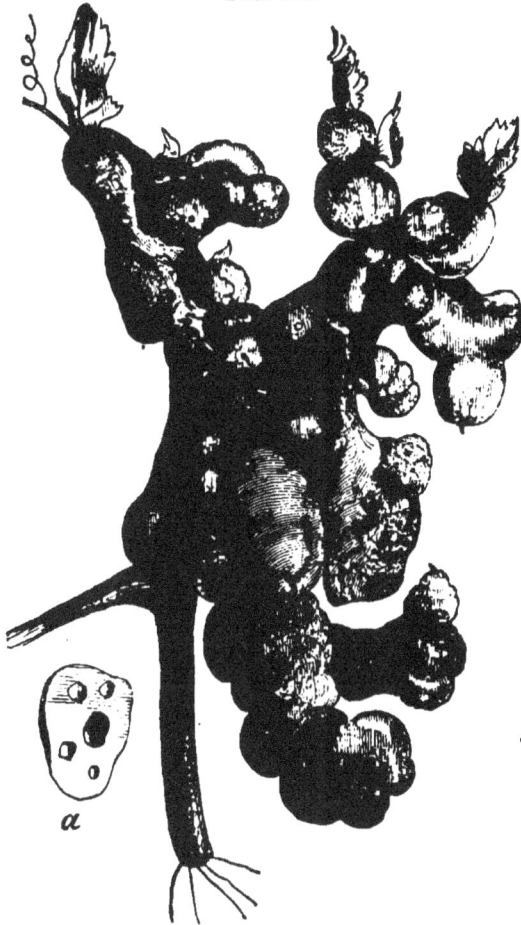

by a species of Thrip, which invade the cells and destroy the inmates.

No. 169.—The Grape-vine Apple-gall.

Vitis pomum Walsh & Riley.

This is a globular, fleshy, greenish gall, about nine-tenths of an inch in diameter, which is attached by a rough base, like that of a hazel-nut, to the stem of the vine. On its external surface there are longitudinal depressions, which divide the gall into eight or nine segments. The interior is fleshy for about one-eighth of its diameter, then follow a series of elongated cells, each divided into two by a transverse partition, the lower being the shorter of the two. Fig. 307, *a*, represents the exterior of the gall; *b*, a section of the same, showing its interior structure. Each cell is occupied by a single larva of a bright-yellow color, with a chestnut-brown, Y-shaped breast-bone,

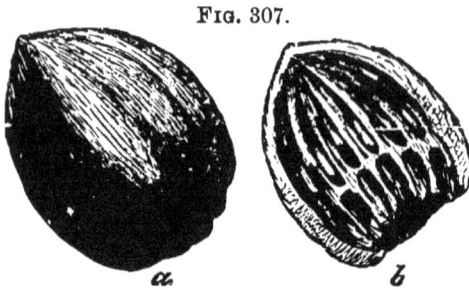

FIG. 307.

which eventually produces a gall-fly belonging to the genus *Cecidomyia*.

This gall sometimes varies in form, being occasionally flattened or depressed; when young it is downy on the outside, succulent within, and is said to have a pleasant, acidulous flavor.

Should any of the galls described ever become a source of annoyance, they may readily be destroyed by hand-picking.

ATTACKING THE FRUIT.

No. 170.—The Grape-seed Insect.

Isosoma vitis Saunders.

This insect was first observed in 1868, when it threatened to become a very troublesome enemy to grape-culture; it was widely distributed, and, having the fecundity usually characteristic of insect life, it might have been expected to increase immensely; but this happily has not been the case, and of late it has seldom prevailed to any serious extent.

About the middle of August some berries in the bunches of grapes may be seen shrivelling up; on opening these, many of them will be found to contain only one seed, and that of an unusually large size; other larger berries will contain two seeds, also swollen, most of the seeds having a dark spot somewhere on their surface. On cutting open these seeds, the kernel will be found almost entirely consumed, and the cavity occupied by a small, milk-white, footless grub, with a pair of brown, hooked jaws, a smooth and glossy skin, with a few very fine, short, white hairs.

Fig. 308 shows this larva highly magnified; the small figure beneath indicates its natural size.

FIG. 308.

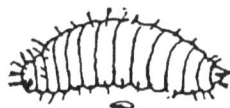

The larva changes to a pupa within the seed during the spring months, and in July emerges as a fly, escaping through a small, irregular hole.

The fly so much resembles that shown in Fig. 309 (which represents a closely-allied form belonging to the same genus) that it is difficult to distinguish between the two; *a* represents the female, *b* the male, *c* the antenna of the female, *d* that of the male, *e* the abdomen of the female, showing the segments or rings of the body, *f* that of the male. All these figures are highly magnified; the short hair-lines underneath the flies indicate the

FIG. 309.

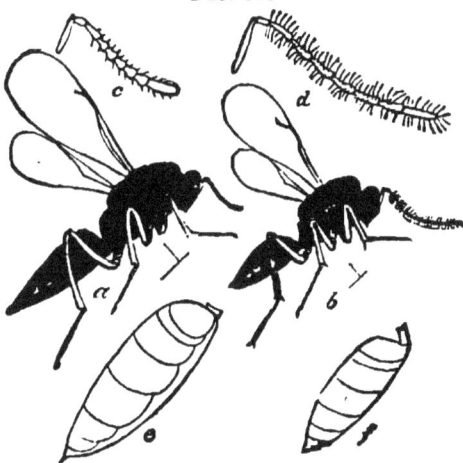

natural size. The fly is black; the head and thorax are finely punctated with minute dots; the abdomen is long and smooth, with a polished surface, and is placed on a short pedicel. The parent insect probably deposits her eggs on the skin of the grape, and the young larvæ, as soon as hatched, puncture the skin and work their way to the seed, which they enter while it is young and soft. Many of the affected grapes have a small scar on their surface, which may indicate where the insect has entered.

Should this tiny foe ever become so troublesome as to require a remedy, the best one suggested is that of carefully gathering and destroying the shrivelled fruit.

No. 171.—The Grape-berry Moth.
Eudemis botrana (Schiff.).

This insect is an imported species, and has long been injurious to grape-culture in the south of Europe. The exact period of its introduction to America is not known, and it is only within the past few years that attention has been called to its ravages. When abundant, it is very destructive; in some instances it is said to have destroyed nearly fifty per cent. of the crop.

The young larvæ are found injuring the grapes early in July, when the infested fruit shows a discolored spot where the larva has entered. (See Fig. 310, c.) When the grape

FIG. 310.

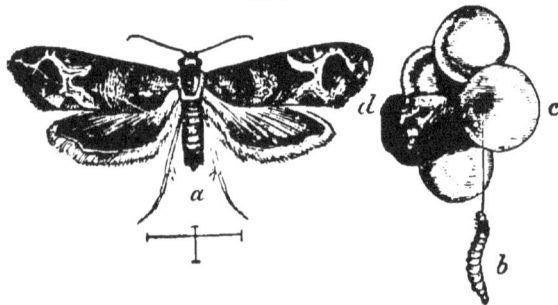

is opened and the contents carefully examined, there will generally be found in the pulp a small larva, rather long and thin, and of a whitish-green color. Besides feeding on the pulp, it sometimes eats portions of the seeds, and if the contents of a single berry are not sufficient, two, three, or more are drawn together, as shown in the figure, and fastened with a patch of silk mixed with castings, when the larva travels from one berry to another, eating into them and devouring their juicy contents. At this period its length is about an eighth of an inch or more; the head is black, and the next segment has a blackish shield covering most of its upper portion; the body is dull whitish or yellowish green. As it approaches maturity, it becomes darker in color, and when about one-fourth of an inch long is full grown. (See *b*, Fig. 310.) The body is then dull green, with a reddish tinge, and a few short hairs, head yellowish green, shield on next segment dark brown, feet blackish, prolegs green.

When the larva is full grown, it is said to form its cocoon on the leaves of the vine, cutting out for this purpose an oval flap, which is turned back on the leaf, forming a snug enclosure, which it lines with silk; frequently it contents itself with rolling over a piece of the edge of the leaf, and within this retreat the change to a chrysalis takes place. The chrysalis is about one-fifth of an inch long, and of a yellow or yellowish-brown color.

The perfect insect, which is shown magnified at *a*, Fig. 310, measures, when its wings are spread, nearly four-tenths of an inch across. The fore wings are of a pale dull-bluish shade, with a slight metallic lustre, becoming lighter on the interior and posterior portions, and are ornamented with dark-brown bands and spots. The hind wings are dull brown, deeper in color towards the margin, the body greenish brown. There are two broods of this insect during the year; the spring brood feed on the tender shoots of the common ironweed (*Vernonia noveboracensis*), also on the tulip-tree.

Remedies.—As it is probable that most of the late brood pass the winter in the chrysalis state attached to the leaves, if these were gathered and burned a large number of the insects would perish. The infested grapes might also be gathered and destroyed. This insect is attacked by a small parasite, which doubtless does its part towards keeping the enemy in subjection.

No. 172.—The Grape Curculio.

Craponius inæqualis (Say).

This is a small beetle belonging to the family of Curculios, which passes the winter probably in the perfect state, and lays its eggs on the young grapes some time in June or early in July. It is a diminutive and inconspicuous insect, only about one-tenth of an inch long. (See Fig. 311, where it is shown

Fig. 811.

Fig. 312.

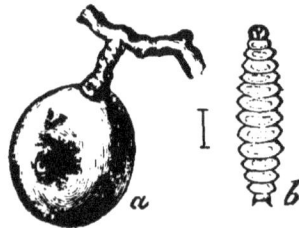

much magnified.) Its color is black, sprinkled with grayish spots and dots, and thickly punctated.

The young larva, when hatched, enters the fruit and begins to feed upon it, its presence being indicated by a discoloration on one side of the berry, as if it were prematurely ripening. A dark, circular dot soon appears in the middle of the colored spot, showing the point where the insect has entered the fruit. The affected berry does not decay, but remains sound and plump; but it sometimes drops to the ground before it is fully ripe. In Fig. 312 a specimen of the injured fruit is shown at *a*, and at *b* a magnified view of the larva, which is an elongated, footless grub, tapering towards the head, about one-fifth of an inch long, the head large, brownish yellow, and horny, the body yellowish white and transparent. Late in July or early in August the larva becomes full grown, when it leaves the berry, drops to the ground, and, burying itself in the soil, changes to a pupa, from which the beetle escapes late in August or early in September.

This is not a common insect, nor is it very generally distributed, and the injury supposed to be done by it to the fruit is often more correctly chargeable to the species last referred to, since that is a much commoner insect. The grape curculio has been observed chiefly in the valley of the Mississippi, but is rarely injurious to any considerable extent or over any large area. Where it is troublesome, the vines may be jarred occasionally during the month of June, placing a sheet or an inverted umbrella under them, when the beetles will fall, and can then be gathered and destroyed, as in the case of the plum curculio.

No. 173.—The Honey Bee.

Apis mellifica Linn.

This useful insect, so valuable to man, is said to have the pernicious habit of puncturing or abrading the skin of the grape and extracting its juices. That the injury thus done is entirely due to the agency of bees has been disputed, some bee-lovers claiming that the grapes are first punctured by birds or bitten by wasps and hornets, and that the bees follow and promptly avail themselves of the store of sweets thus laid open for their use. The evidence, however, on the whole, seems rather strong against the bees, and there is little doubt that they frequently do abrade the skin of the fruit with their claws and afterwards extract the sweets with their brush-like tongue.

SUPPLEMENTARY LIST OF INJURIOUS INSECTS WHICH AFFECT THE GRAPE.

ATTACKING THE CANES.

The apple-twig borer, No. 13, the tree cricket, No. 178,

and the mealy flata, No. 218, all injure the canes of the grape.

ATTACKING THE LEAVES.

The fall web-worm, No. 27 ; the saddle-back caterpillar, No. 49 ; and the smeared dagger, No. 194.

ATTACKING THE FRUIT.

The Indian Cetonia, No. 81.

INSECTS INJURIOUS TO THE RASPBERRY.

ATTACKING THE ROOTS.

No. 174.—The Raspberry Root-borer.

Bembecia marginata Harris.

This borer is quite distinct from the cane-borer, No. 176, that insect being without legs in the larval state, while this one has sixteen legs, a feature which will enable any person readily to distinguish the one from the other. The raspberry root-borer belongs to the same family of clear-winged moths as the peach-borer, and there is a striking resemblance between the two species in the several stages of their existence.

Both the male and the female moth are shown in Fig. 313, where *a* represents the male, and *b* the female. The front wings are transparent, veined with black or brownish, and heavily margined with reddish brown; the hind wings are transparent, with dark veins, and both wings are fringed with dark brown. The body is black, prettily banded and marked with golden yellow, as in the figure. The wings, when expanded, will measure from three-quarters of an inch to an inch across.

The eggs are said to be deposited by the female during the hot summer weather on the leaves of the raspberry, and the young larva, when hatched, finds its way from these to the stem or cane, and there feeds upon the pithy substance in the interior,

FIG. 313.

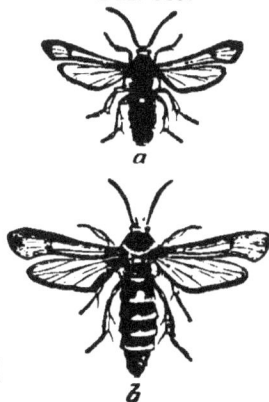

303

and gradually channels the cane to the root, in which it spends the winter months, forming before spring cavities of considerable extent. As the spring opens, it works its way up again, usually through the interior of another cane, to a height of five or six inches, where the larva, in preparing for the exit of the future moth, cuts the cane in one place nearly through, leaving a mere film of skin unbroken. When full grown, it is about an inch long, of a pale-yellow color, with a dark-brown head, and a few shining dots on each segment of the body. Within the cane, and near the spot specially prepared by the larva, the change to a chrysalis takes place, and when the time approaches for the moth to escape, the chrysalis wriggles itself forward, and, pushing against the thin skin remaining on the cane, ruptures it, and, forcing its way through the opening, there awaits the escape of the moth, which usually takes place within a few hours afterwards.

The injury thus done to the root is often followed by the death of the canes, a result sometimes incorrectly attributed to the severe cold of winter. Little can be done towards the destruction of this pest other than by laying bare the roots and cutting out the infested portions. A parasitic insect is said to attack these root-borers, and probably destroys many of them.

FIG. 314.

No. 175.—The Raspberry-root Gall-fly.

Rhodites radicum Osten Sacken.

This is a small gall-fly, which produces a large brown gall on the roots, a good representation of which is given in Fig. 314. The swelling is composed of a yellow, pithy substance, scattered throughout which are a number of cells, each enclosing a small white larva, the progeny of the

gall-fly. These soon change to pupæ, and they in turn produce after a time the perfect insects, which eat their way
out through the substance of the gall, leaving small holes to
mark the place of exit. These galls are not only the abode
of the makers, the gall-flies, but are also frequented by other
species known as guest-flies, and the presence of these as well
as other parasitic species in company with the normal inmates
is apt to perplex the observer, and renders it more difficult to
discover the real authors of the mischief. This gall chiefly
affects the black raspberry ; it also occurs on the blackberry,
and sometimes on the roots of the rose.

Wherever these excrescences are found they should be collected and burnt.

ATTACKING THE CANES.

No. 176.—The Raspberry Cane-borer.

Oberea bimaculata Oliv.

This insect in the larval state lives in the centre of the
cane, where it burrows a passage from above downwards,
often causing the death of the cane. Its natural home is
among the wild raspberries, but it has taken very kindly to
the cultivated sorts, and appears indeed to prefer them.

The perfect insect is a long-horned beetle (see Fig. 315),
with a long and narrow black body, with the top of the
thorax and the fore part of the breast pale yellowish ; the wing-cases are covered with coarse   FIG. 315.
indentations and slightly notched at the ends, and
there are two black spots on the thorax, which,
however, are sometimes wanting, and a third black
dot on the hinder edge, just where the wing-covers
join the thorax.   The beetles appear on the wing during the
month of June, and, after pairing, the female proceeds to
deposit her eggs, which she does in a very singular manner.

With her mandibles she girdles the young growing cane near the tip in two places, one ring being about an inch below the other, and between the rings the cane is pierced, and an egg thrust into its substance near the middle, its location being indicated by a small, dark-colored spot. The supply of sap being impeded or stopped, the tip of the cane above the upper ring soon begins to droop and wither, and shortly dies, when a touch will sever it at the point where it has been girdled.

The egg is long and narrow and of a yellow color, is quite large for the size of the insect, and, embedded in the moist substance of the cane, absorbs moisture and increases in size until in a few days a small grub hatches from it. The larva as it escapes from the egg is about one-fourteenth of an inch long, with a yellow, smooth, glossy body, roughened at the sides, and clothed with very minute short hairs. The head is small and reddish brown, and the anterior segments of the body swollen; it is also footless. The young larva burrows down the centre of the stem, consuming the pith until full grown, which is usually about the end of August, when it is nearly an inch long and of a dull-yellow color, with a small, dark-brown head. By this time it has eaten its way a considerable distance down the cane, in which it remains during the winter, and where it changes to a pupa, the beetle escaping the following June, when it gains its liberty by gnawing a passage through. This borer injures the blackberry as well as the raspberry.

The presence of these enemies is easily detected by the sudden drooping and withering of the tips of the canes. They begin to operate late in June, and continue their work for several weeks; hence by looking over the raspberry plantation occasionally at this season of the year and removing all the withered tops *down to the lowest ring,* so as to insure the removal of the egg, these insects may be easily kept under, for they are seldom numerous.

No. 177.—The Red-necked Agrilus.

Agrilus ruficollis (Fabr.).

In the spring-time, when raspberry and blackberry canes are being pruned, they will often be observed swollen in places to the length of an inch or more, in the manner shown in Fig. 316. This swelling is a pithy gall, and has been named the Raspberry Gouty-gall, *Rubi podagra* Riley, and is produced by the irritation caused by the presence of the larva of the red-necked Agrilus. The swollen portions are not smooth, as the healthy ones are, but have the surface roughened with numerous brownish slits and ridges, and when the ridges are cut into with a knife, there will be found under each of them the passage-way of a minute borer, and either in the channel or in the soft substance adjoining, the larva will usually be found. Fig. 317 represents the nearly full-grown larva magnified, the hair-line at the side indicating its natural size. Its body is almost thread-

FIG. 316.

FIG. 317.

like, and of a pale-yellowish or whitish color, with the anterior segments enlarged and flattened. The head is small and brown, the jaws black, and the tail is armed with two slender, dark-brown horns, each having three blunt teeth on the inner edge. When full grown, it measures about six-tenths of an inch long. While young it inhabits chiefly the sap-wood, and, following an irregular, spiral course, frequently girdles and destroys the cane; usually several larvæ will be found in the one cane, thus lengthening the gall and causing it to assume a very irregular shape. In April or May the larva penetrates into the pith, where it is more secure from insect and other foes, and there changes to a pupa, from which the perfect beetle escapes early in the summer.

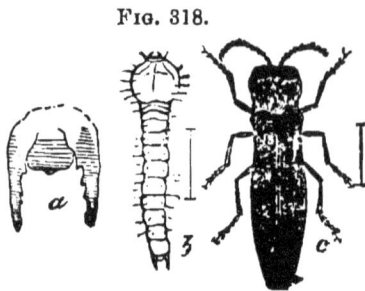

FIG. 318.

The eggs are deposited on the young canes probably in July, and the tiny young larvæ, when hatched, eat into the cane, producing, in time, the mischievous results already detailed. Fig. 318, *c*, shows the perfect insect, magnified; *b*, another view of the larva, and *a* the horns at the end of its body, much magnified. The beetle is about three-tenths of an inch long, with a rather small, dark bronzy head, a beautifully bright coppery neck, and brownish-black wing-covers. The under surface is of a uniform shining black color.

The best method of destroying this insect is to cut out the infested canes in the spring and burn them before the beetle escapes.

No. 178.—The Tree Cricket.

Œcanthus niveus Serv.

Of all the insects affecting the canes of the raspberry, probably this is the most troublesome. Fig. 319 represents the male, and Fig. 320 the female. They are about seven-tenths of an inch long, of a pale whitish-green color, and semi-transparent, with several dusky stripes on the head and thorax; the legs and antennæ are also dusky or dark-colored. They are exceedingly lively, and the males quite musical, chirping merrily with a loud, shrill note among the bushes all the day. In the autumn they attain full growth, and it is then that the female,

FIG. 319.

in carrying out her instinctive desires to protect her progeny, becomes such an enemy to the raspberry-grower. She is furnished with a long ovipositor, which she thrusts obliquely more than half-way through the cane, and down the open-

FIG. 320.

ing thus made she places one of her eggs, which are yellowish and semi-transparent, about one-eighth of an inch long, and narrow; a second one is then placed, in the same manner, alongside of the first, and so on, until from five to fifteen eggs have been placed in a row. In Fig. 321 is shown a piece of infested cane; *a* represents the irregular row of punctures indicating the presence of the eggs; *b*, the same laid open, showing the eggs in position; at *c* is a magnified egg, while *d* shows the granulated head of the same, still more highly magnified. Owing to the presence of these eggs, the cane is much weakened, and is liable to break on slight provocation; sometimes the part beyond the punctures dies, but if it survives, and escapes being broken in winter, it is very apt to break from the action of the wind on the weight of foliage as soon as it has expanded in spring, and the crop which would otherwise be realized is lost.

FIG. 321.

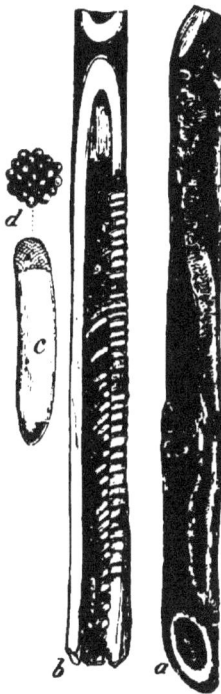

As soon as the spring opens, the eggs begin to swell, and about midsummer, or sometimes a little earlier, the young insects hatch, which much resemble the perfect insect in form, but lack wings. They at once leave the raspberry canes and do no further injury to them. At first they feed more or less on plant-lice, and later in the season on ripe fruits and other succulent food. Besides injuring the raspberry and blackberry, they attack the canes of the grape and the smaller branches of plum, peach, and other trees.

Remedies.—Cut out late in the fall or early in the spring all those portions of the cane which contain eggs, and burn them. Wherever the eggs are deposited the regular rows of punctures are easily seen, and often their presence is rendered still more apparent by a partial splitting of the cane. The mature insects may also be destroyed in the autumn by suddenly jarring the bushes or canes on which they collect, when they drop to the ground, and may be trodden under foot before they have time to hop or fly away.

ATTACKING THE FLOWERS.

No. 179.—The Pale-brown Byturus.

Byturus unicolor Say.

This insect is a small beetle, which is sometimes very destructive to the blossoms of the raspberry. It is a native insect, about three-twentieths of an inch long, of a yellowish-brown or pale-reddish color, and densely covered with fine, pale-yellow hairs. The surface of the body, when seen under a magnifying-lens, is densely punctated. This beetle is shown, both magnified and of the natural size, in Fig. 322.

Fig. 322.

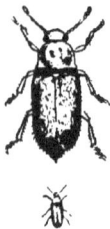

Late in May and early in June, when the flowers are expanding, this insect is busily employed eating into and injuring or destroying the flower-buds. At this period many of the flower-buds may be found with a hole in the side, through which the enemy has entered and eaten away, partly or wholly, the stamens, also the spongy receptacle on which they are borne. Where the injury is only partial, the flower usually expands; but when the sexual organs are entirely destroyed, as is often the case, the buds generally wither and do

not open. The beetles attack the expanded flowers as well
as those which are unopened, partly hiding themselves about
the base of the numerous stamens on which they are feeding.
They are seldom seen during the middle of the day, but work
chiefly during the early hours of the morning and evening.
They feed on the blossoms of the blackberry also, and are
said to eat the leaves of the raspberry occasionally.

Where the flowers are injured, the fruit, if it forms at all,
is always imperfect; hence, should this insect become very
plentiful, it would prove a great hindrance to successful rasp-
berry-culture. Fortunately, it has never yet occurred in any
great numbers; should it at any time become numerous, its
ranks might be thinned by hand-picking.

ATTACKING THE LEAVES.

No. 180.—The Raspberry Saw-fly.

Selandria rubi Harris.

The perfect insect in this instance is a four-winged fly belonging to the order *Hymenoptera,* which appears from about
the 10th of May to the beginning of June, or soon after the
young leaves of the raspberry are put forth. Fig. 323 gives a
magnified view of this fly.
The wings, which are transparent, with a glossy surface
and metallic hue, measure,
when expanded, about half
an inch across; the veins
are black, and there is also
a streak of black along the
front margin, extending
more than half-way to-
wards the tip of the wing. The anterior part of the body is
black, the abdomen dark reddish. In the cool of the morning,

FIG. 323.

when these flies are approached as they rest on the bushes, they have the habit of falling to the ground, and there remaining inactive long enough to permit of their being caught; but with the increasing heat of the day they become much quicker in their movements, and take wing readily when approached.

The eggs are buried beneath the skin of the raspberry leaf, near the ribs and veins, and are placed there by means of the saw-like apparatus with which the female is provided. The egg is white and semi-transparent, with a faint yellow tinge, and a smooth, glossy surface, oval in form, and about one-thirtieth of an inch long. The skin covering it is so thin and transparent that the movements of the enclosed larva may be observed a day or two before it is hatched, and the black spots on the sides of the head are distinctly visible; it escapes through an irregular hole made on one side of the egg.

The newly-hatched larva is about one-twelfth of an inch long, with a large, greenish-white head, having a black, eye-like spot on each side; the body nearly white, semi-transparent, and thickly covered with transverse rows of white spines. As it grows older it becomes green, very much the color of the leaf on which it is feed-

FIG. 324.

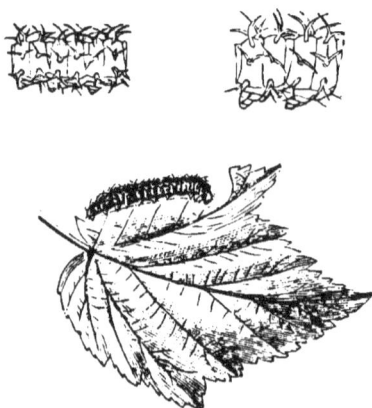

ing, and on this account it would be difficult to discover were it not that it riddles the leaves by eating out all the soft tissues between the coarser veins. When full grown, it measures about three-quarters of an inch in length, is of a dark-green color, its body thickly set with pale-green, branching tubercles. The head is small, pale yellowish green, with a dark-brown dot on each side. This larva is usually found on the upper surface of the leaf. In

Fig. 324 it is shown of the natural size, with some of the segments magnified, showing the arrangement of the spines on the back and side.

On reaching maturity, which is usually from the middle to the end of June, the larva leaves the bush, and, descending to the ground, penetrates beneath the surface, and there constructs a little, oval, earthy cocoon, mixed with silky and glutinous matter. These cocoons are toughly made, and may be taken out of the earth in which they are embedded, and even handled roughly, without much danger of dislodging the larvæ. They remain within the cocoon for a considerable time unchanged, finally transforming to pupæ, from which the flies escape early the following spring.

These insects may be readily destroyed by syringing or sprinkling the bushes with water in which powdered hellebore has been mixed, in the proportion of an ounce of the powder to a pailful of water.

No. 181.—The Raspberry Apatela.

Apatela brumosa (Guen.)

The caterpillar of this moth, although never yet recorded as very injurious, is more or less common on raspberry bushes every year in some localities. It does not appear in flocks, but feeds singly. It is a gray hairy caterpillar, which attains full growth during the latter part of July or in August, when it measures, if in motion, about an inch and a quarter long, but when at rest, owing to some of the segments of the body being drawn partly within the others, it does not measure more than an inch. The body is thickest from the third to the seventh segment, tapering a little anteriorly and posteriorly, and is of a brownish-black color, with a transverse row of paler tubercles on each segment, from which spring clusters of brownish-white or grayish hairs of varying lengths. Behind the third segment there is a space down the centre of the back where the dark color of the body is distinctly seen. The head is of a shining black color, the upper portion overhung by the long hairs of the next segment. The under side is greenish brown, with a few small clusters of short brown hairs.

The larva changes to a brown chrysalis within a rather tough cocoon formed of pieces of leaves interwoven with silk.

The moth (Fig. 325) has the fore wings gray, mottled with spots, streaks, and dots of darker shades of gray and brown.

Fig. 325.

The hind wings are of a dull pale gray, deepening in color a little towards the outer margin. The under surface is paler than the upper. When the wings are expanded, they measure about an inch and a quarter across.

Should this insect ever become troublesome, it may be subdued by hand-picking, or destroyed by showering the bushes with water in which hellebore or Paris-green has been mixed, in the proportion of an ounce of the former or one or two teaspoonfuls of the latter to two gallons of water.

No. 182.—The Raspberry Plume-moth.

Oxyptilus nigrociliatus Zeller.

The caterpillar of this pretty little plume-moth has not in any instance on record been sufficiently numerous to

Fig. 326.

be considered destructive, yet it is an interesting insect, and on this account deserves a passing notice. About the middle of June the larva reaches full growth, when it is about four-tenths of an inch long, of a pale yellowish-green color, streaked with pale yellow, and with transverse rows of shining tubercles, from each of which arise from two to six spreading hairs of a yellowish-green color. The head is small, pale green, with a faint brown dot on each side. Fig. 326 represents this larva, much magnified.

When the larva is about to change to a chrysalis, it spins a loose web of silk on a leaf or other suitable spot, to which the chrysalis is attached. This is less than three-tenths of an inch long, pointed behind, enlarging gradually towards the front, where, near the end, it slopes abruptly to the tip. Its color is pale green, with a line along the back of a deeper shade, margined on each side with a whitish ridge; it is also more or less hairy. In about a week or ten days the chrysalis changes to a darker color, shortly after which the perfect insect escapes.

The moth (Fig. 327), although quite small, is very beautiful; it measures, when its wings are expanded, about half an inch across. The fore wings are of a deep brownish-copper color, with a metallic lustre, and a few dots of silvery white; they are cleft down the middle about half their depth, the division as well as the outer edge being fringed. The hind wings, which resemble the fore wings in color, are divided into three portions, the hinder one being almost linear, and all deeply fringed. The antennæ are ringed with silvery white, and there are spots of the same color on the legs and body.

Fig. 327.

Should this insect at any time prove troublesome, it might be easily destroyed with powdered hellebore and water, as recommended for No. 181.

No. 183.—*Chelymorpha Argus* Leichtenstein, a beetle belonging to the family *Chrysomelidæ,* is also said to feed occasionally on the raspberry. In Fig. 328 the beetle is represented of the natural size, the pupa in Fig. 329. It can scarcely be regarded as injurious, and needs but a passing notice.

Fig. 328.

Fig. 329.

ATTACKING THE FRUIT.

No. 184.—The Raspberry Geometer.

Synchlora rubivoraria (Riley).

The larva of this pretty moth feeds chiefly on the fruit of the raspberry; it is said that it occasionally feeds also on the leaf.   Fig. 330 shows the larva, of natural size, on the fruit

FIG. 330.

at *a*; *b*, an enlarged view of one of the segments of its body, showing the hairs with which it is adorned.   The moth, of the natural size, is seen at *c*, while at *d* an enlarged outline is given of one pair of the wings.

The larva reaches maturity about the time of the ripening of the raspberry, when it is about three-quarters of an inch long, of a yellowish-gray color, each segment being furnished with several short prickles.   It has the habit of disguising itself by attaching to its thorny projections tiny bits of vegetable matter, such as the anthers of flowers, bits of leaves, etc., and by this means it often escapes detection.

When full grown, the larva forms a slight cocoon, within which it changes to a chrysalis of a pale-yellow color, with darker lines and spots, which in a few days produces the perfect insect.

The wings of the moth are of a delicate pale-green color, crossed by two lines of a lighter shade, and, when expanded, they measure about half an inch across. The body is green above and white beneath.

As the larva of this insect is not usually observed until the fruit is ripe, no poisonous applications to destroy it could be used, and resort must be had, if anything is done, to hand-picking. One species of parasitic insect is known to prey on it.

No. 185.—The Flea-like Negro-bug.

Corimelæna pulicaria Germ.

This disgusting little pest is not at all uncommon on ripe raspberries. Its presence may be discovered by the fruit having a nauseous *buggy* odor, but the insect is so small that it is often taken into the mouth un-noticed until the disgusting flavor reveals its presence. In Fig. 331 we have a magnified outline of this insect, the smaller sketch at the side showing its natural size. It is of a black color, with a whitish stripe along each side, and is furnished with a pointed beak or sucker, with which it punctures the fruit and extracts its juices. This troublesome visitor is also found on the blackberry, and occasionally on the strawberry.

Fig. 331.

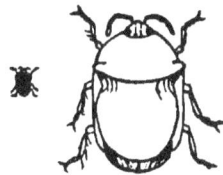

SUPPLEMENTARY LIST OF INJURIOUS INSECTS WHICH AFFECT THE RASPBERRY.

ATTACKING THE LEAVES.

The fall web-worm, No. 27 ; the oblique-banded leaf-roller, No. 35 ; the saddle-back caterpillar, No. 49 ; the apple leaf-miner, No. 50 ; the yellow woolly-bear, No. 146 ; the pyramidal grape-vine caterpillar, No. 147 ; the neat strawberry leaf-roller, No. 193 ; the smeared dagger, No. 194 ; and the cucumber flea-beetle, No. 223.

ATTACKING THE CANES.

No. 186.—The Pithy Gall of the Blackberry.

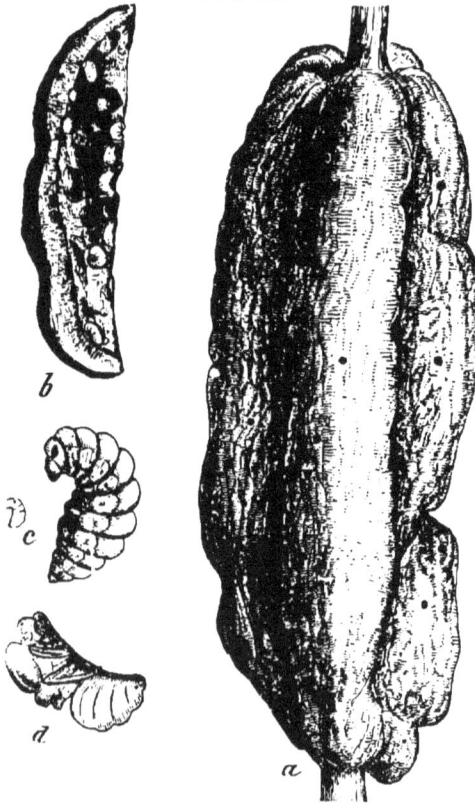

This curious gall, which is represented in Fig. 332, is sometimes found on blackberry canes. It is about two or three inches long, of a dark-red or reddish-brown color, oblong in form, with its surface uneven, with deep longitudinal furrows, which divide the gall more or less completely into four or five portions. It is caused by a small four-winged fly, *Diastrophus nebulosus* Osten Sacken. If a transverse section of this gall be made, there will be found about the middle a number of oblong cells about one-eighth of an inch long, shown at *b* in the figure, each containing a single larva or pupa. The larva, which is represented enlarged at *c*, is about one-tenth of an inch long, white, with the mouth parts

Fig. 332.

318

reddish, and the breathing-pores and an oval spot on each side behind the head of the same color. The insect usually remains in the larval state during the greater part of the winter, then changes to a pupa (*d*, Fig. 333), the perfect insect appearing in spring. The fly is about one-twelfth of an inch long, black, with transparent wings and red feet and antennæ.

These gall-makers are attacked by parasitic insects, and are also devoured by birds.

No. 187.—The Seed-like Gall of the Blackberry.

This is a singular gall, about one-tenth of an inch in diameter, which sometimes occurs in clusters around the canes of the blackberry, covering them with a belt of these seed-like bodies to the depth of an inch or an inch and a half. They are round, of a reddish color, and from many of them arise more or less strong spines, and when cut into, unless they have already been emptied by birds, each one will be found to contain a single larva or pupa. These galls are also caused by a small, four-winged fly closely related to that of the pithy gall, and known as *Diastrophus cuscutæformis* Osten Sacken. It is of a dark-brown or black color, with red feet and antennæ.

No. 188.—The Blackberry Bark-louse.

Lecanium ———?

An undetermined species of Lecanium is sometimes found on the canes of the blackberry. This louse is of an irregular hemispherical form, about one-fourth of an inch in diameter, and of a shining mahogany color. It appears in groups or masses attached to the canes, and each one, when lifted, is found to cover a large number of pale-pinkish eggs. This is very similar to the grape-vine bark-louse, No. 126, and may be treated in the same manner.

ATTACKING THE LEAVES.

No. 189.—The Blackberry Flea-louse.

Psylla rubi W. & R.

This insect has been reported as common on blackberry leaves in some parts of New Jersey. ˙It is a small, four-winged fly, much resembling the pear-tree Psylla (No. 70), about one-eighth of an inch long when its wings are closed. The mature insect is like a plant-louse in appearance, but its transparent wings are differently veined, and it has the power of jumping briskly when disturbed, which plant-lice never possess. The leaves affected curl up so as to make a safe harbor for the lice-like larvæ, which occupy these enclosures during the greater part of the summer. To lessen their numbers, gather the curled leaves and burn them.

SUPPLEMENTARY LIST OF INJURIOUS INSECTS WHICH AFFECT THE BLACKBERRY.

ATTACKING THE ROOTS.

The raspberry root-borer, No. 174, and the raspberry-root gall-fly, No. 175, both injure the roots of the blackberry.

ATTACKING THE CANES.

The raspberry cane-borer, No. 176, and the red-necked Agrilus, No. 177.

ATTACKING THE LEAVES.

The fall web-worm, No. 27; the apple leaf-miner, No. 50; the waved Lagoa, No. 89; the yellow woolly-bear, No. 146; and the neat strawberry leaf-roller, No. 193.

ATTACKING THE FRUIT.

The flea-like negro-bug, No. 185, is common on the fruit.

INSECTS INJURIOUS TO THE STRAWBERRY.

ATTACKING THE ROOTS.

No. 190.—The Strawberry Root-borer.

Anarsia lineatella Zeller.

When occurring in great numbers, this insect is very injurious, playing sad havoc with the strawberry-plants. The borer is a small caterpillar, nearly half an inch long, and of a reddish-pink color, fading into dull yellow on the second and third segments, the anterior portion of the second segment above being smooth, horny-looking, and brownish yellow like the head. On each segment there are a few shining, reddish dots, from every one of which arises a single, fine, yellowish hair. The under surface is paler. This borer eats irregular channels through the crown, sometimes excavating large chambers, at other times tunnelling it in various directions, eating its way here and there to the surface. If examined in the spring, most of the cavities will be found to contain a moderate-sized, soft, silky case, nearly full of castings, which doubtless has served as a place of retreat for the larva during the winter.

Early in June, when mature, the caterpillar changes to a small, reddish-brown chrysalis, either within one of the cavities excavated in the crown, or among decayed leaves or rubbish about the surface, from which the moth escapes early in July.

The moth (see Fig. 333) is very small, of a dark-gray color, with a few blackish-brown spots and streaks on the fore wings. The fringes bordering the wings are gray tinged with yellow. The moth lays an egg on the crown of the plant late in July or early in August, which soon hatches; the small caterpillar burrows into the heart of the plant, and remains in one of the chambers during the winter, occupying one of the silky cases referred to. The channels formed by this larva through the crown and larger roots of the plant soon cause it to wither and die; or, if it survives, to send up weakened and almost barren shoots.

This insect does not limit its depredations to the strawberry; the larva is also found boring into the tender twigs of the peach-tree and killing the terminal buds.

FIG. 333.

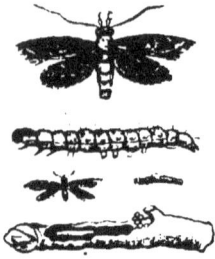

In Fig. 333 we have a representation of the larva and moth, both of the natural size and magnified, also of an injured peach-twig. The insect is known to attack the peach-tree in Europe, whence it has probably been imported to this country.

Remedies.—Dusting the plants with air-slaked lime or with soot has been recommended, but there seems to be no way thoroughly to destroy this pest except by digging up the strawberry plants, burning them, and planting afresh. The larvæ are subject to the attacks of parasites, which doubtless materially limit their increase.

No. 191.—The Strawberry Crown-borer.

Tyloderma fragariæ (Riley).

This is an indigenous insect, a beetle belonging to the family of Curculios. The beetle (Fig. 334) appears in June or July, and deposits an egg about the crown of the plant, from which, when hatched, the larva burrows downwards, eating into the substance of the crown. Here it remains, boring and excavating, until it attains full growth, when it appears as shown at *a* in the figure, where it is much magnified. It is about one-fifth of

FIG. 334.

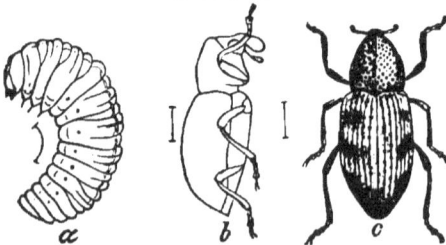

an inch long, white, with a horny, yellow head. It changes to a pupa within the root, from which the beetle escapes during the month of August.

The beetle, shown at *b* and *c* in the figure, is about one-sixth of an inch long, of a brown color, with several more or less distinct dark-brown spots, and is marked with lines and dots.

Almost all the plants infested with this larva are sure to perish, and old beds are said to be more liable to injury than new ones. The only remedy suggested is to dig up and burn the plants after the fruiting season is over, and before the larva has time to pass through its transformation and escape as a beetle.

ATTACKING THE LEAVES.

No. 192.—The Strawberry Leaf-roller.

Phoxopteris comptana Frol.

This insect, which is sometimes designated *the* strawberry leaf-roller, is not the only leaf-roller which attacks the leaves of the strawberry. The caterpillars belonging to the early brood are found upon the plants during the month of June, rolling the leaves into cylindrical cases, fastening them with threads of silk, and feeding within on their pulpy substance, causing the leaves to appear discolored and partly withered. They are about one-third of an inch long, and vary in color from yellowish brown to a darker brown or green. The head is yellowish and horny, with a dark eye-like spot on each side. The second segment has a shield above, colored and polished like the head, and on every segment there are a few pale dots, from each of which arises a single hair. In Fig. 335, *a* represents the larva of its natural size, *b* a magnified view of the head and four succeeding segments, and *d* the terminal segment of the body.

The larva becomes a chrysalis within the folded leaf late in

June, and appears as a moth early in July. The fore wings of the moth are reddish brown, streaked and spotted with black and white, as shown in the figure at *c;* the hind wings and abdomen are dusky; the head and thorax reddish brown. When expanded, the wings measure nearly half an inch across. The eggs for the second brood of larvæ are deposited during the latter part of July, the larvæ attaining their full growth towards the end of September, when they change to chrysalids, and remain in that condition during the winter, producing moths the following spring.

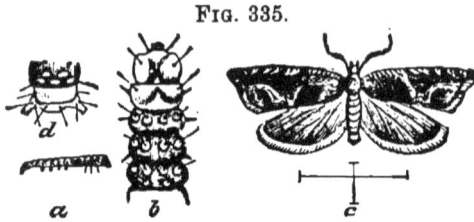

FIG. 335.

This species is sometimes very destructive, when the plants should be sprinkled with a mixture of powdered hellebore and water, in the proportion of an ounce to the pailful, or the rolled leaves may be gathered and burnt, or the plantation ploughed up in the autumn or early in the spring, and the insects destroyed by burying them; in replanting, avoid using plants from infested districts.

No. 193.—The Neat Strawberry Leaf-roller.

Eccopsis permundana (Clemens).

This pernicious little caterpillar appears just about the time that the strawberry blossoms are opening, and delights to form its protecting case by drawing the flowers and flower-buds together into a ball and to feast upon their substance, a peculiarity which renders its attacks much more injurious than any mere consumption of leaves would be. The larva is of a green color, with the head and upper part of the next segment black. When full grown, it is about five-eighths of an inch long, is very active in its habits, and wriggles itself quickly out of its hiding-place when disturbed. Late in

June or early in July it changes to a brown chrysalis, from which, in a few days, the perfect insect escapes.

The moth, which is shown magnified in Fig. 336, has its fore wings yellowish or greenish brown, varying much in shade of color, with irregular, lighter markings crossing the wings obliquely ; the hind wings are ashy brown.

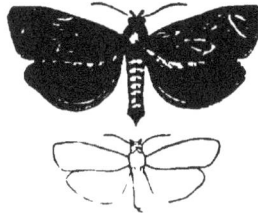

FIG. 336.

The caterpillar is very destructive in some districts, and feeds upon the wild strawberry as well as upon the cultivated varieties; also upon the leaves of the raspberry and blackberry.

Remedies.—Dusting the plants with air-slaked lime, soot, or ashes, or sprinkling them with a mixture of Paris-green and water, in the proportion of one or two teaspoonfuls to two gallons of water, would no doubt prove beneficial. The caterpillar is very subject to the attacks of parasites.

No. 194.—The Smeared Dagger.

Apatela oblinita (Sm. & Abb.).

The moths belonging to the genus Apatela are called "daggers" in England, on account of a peculiar dagger-like mark found on the front wings near the hind angle. This peculiarity being partly obliterated in this species, it has received the common name of the " smeared dagger."

The accompanying figure, 337, represents the insect in its various stages. The larva, *a*, is a hairy caterpillar, brightly ornamented, and about an inch and a quarter long. It is of a deep velvety black color, with a transverse row of tubercles on each segment, those above being bright red and set in a band of the same color, which extends down each side. From each tubercle there arises a tuft of short, stiff hairs, those on the upper part of the body being red, while below they are yellowish or mixed with yellow. On each side of an imaginary line drawn down the centre of the back is a row of bright-yellow spots, two or more on each segment, and below these, and close to the under surface, a bright-yellow band, deeply indented on each segment. Spiracles white. There are also a few whitish dots scattered irregularly over the surface of the body. This caterpillar is so conspicuous for its beauty that it is sure to attract the attention of every beholder.

As soon as it is full grown, it draws together a few leaves

Fig. 337.

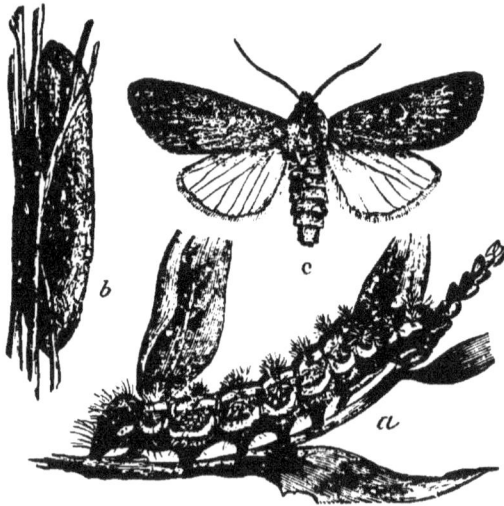

or other loose material, and, with the aid of some silk, constructs a rude case (*b*, Fig. 337), within which it changes to a dark-brown chrysalis. The caterpillars of the fall brood, which become chrysalids early in September, do not produce moths until June following. There are two broods during the season, but the members of the early one, being less abundant, are not so often seen as those of the later brood.

The moth, which is represented at *c* in the figure, is a very plain-looking insect. Its fore wings are gray, with a row of blackish dots along the hind border. A broken, blackish, zigzag line, sometimes indistinct, crosses the wing beyond the middle, and there are some darker grayish spots about the middle of the wing; the hind wings are white.

This caterpillar is not confined to the strawberry, but feeds also on the leaves of the grape, apple, peach, raspberry, willow, and on the common smart-weed, *Polygonum punctatum.* Being such a general feeder, it is never likely to become injurious. It is preyed upon by several parasitic insects, which no doubt render material aid in keeping it within due limits.

No. 195.—Cut-worms.

Under No. 45, among the insects injurious to the apple, the reader will find reference made to those species of cut-worms which are noted for climbing trees and devouring the foliage. These climbing cut-worms eat also anything on the ground which may come in their way. There are, however, a number of species which do not climb trees, and it is, as a rule, among these that we find the greatest enemies to strawberry-plants. These larvæ, or "worms," as they are called, all have a general resemblance to one another, being smooth and of some shade of greenish gray or brown, with dusky markings, or occasionally almost black. Both the larvæ and the moths are nocturnal in their habits, and secrete themselves during the day, the moths in crevices of the bark of trees or other suitable hiding-places, while the larvæ bury themselves under the ground in the neighborhood of the scene of their depredations. Their life-history is briefly told under No. 45, and need not be repeated here. It will suffice in this connection to refer to several representative species of the class which do not climb.

The Greasy Cut-worm, *Agrotis Ypsilon* (Rott.). This larva, which is shown in Fig. 338, is of a deep dull-brown color, inclining to black, with paler longitudinal lines, a faint, broken, yellowish-white line along the back, and two other indistinct pale lines on each side; there are also a few shining black dots on each segment. When full grown, it is about an inch and a half long.

The moth, also represented in the figure, has the fore wings brownish gray with darker markings, and patches of a paler color towards the apex of the wing. The hind wings are almost white, with a pearly lustre, and nearly semi-transparent. When the wings are spread, they measure about an inch and three-quarters across.

FIG. 338.

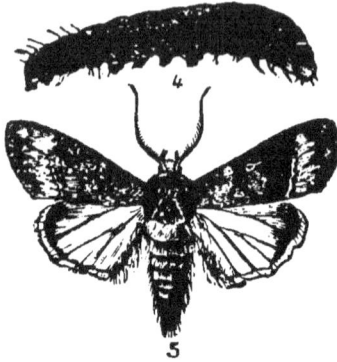

This is one of the most abundant of cut-worms, being found from Georgia and Texas to Nova Scotia and Manitoba, also in Europe, Asia, Africa, and Australia. The caterpillar attacks all sorts of garden products, and is one of the cotton cut-worms of the South.

FIG. 339.

The Striped Cut-worm, *Agrotis tricosa* Lintner. This caterpillar is of an ash-gray color, with broad, dark longitudinal lines, and several narrow lighter ones, and when full grown is nearly an inch and a half long. The moth is shown in Fig. 339 with its wings expanded. The fore wings are of a dark-brown color, paler towards the front edge, with pale-gray markings along the veins. The hind wings are of a dark smoky brown, becoming gradually paler towards the body.

FIG. 340.

The Checkered Rustic, *Agrotis tessellata* Harris (Fig. 340), is of a dark-ash color, with two pale spots on the front wings alternating with a triangular and a nearly square black spot.

The Glassy Cut-worm, *Hadena devastatrix* (Brace). In Fig. 341 we have a representation of the larva. It is of a shining green color, with a red head and a dark-brown, horny-

FIG. 341.

FIG. 342.

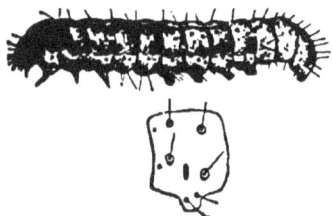

looking shield on the next segment. On each ring there are a number of shining dots, from each of which arises a single short hair, as seen in the magnified segment below. The moth (Fig. 342) is of a dark ashen-gray color, marked with black and white spots, streaks, and dots; the hind wings are pale brownish gray.

Many more examples of these cut-worms and their moths might be cited, but enough has been given to show their general characteristics.

To subdue these insects is no easy matter, since they do not usually eat the foliage in the manner that other caterpillars do, but attack the plant at about the base, and, having cut it through, leave the greater portion of it to wilt and perish. Sprinkling the plants with air-slaked lime, ashes, or powdered hellebore, or showering them well with water containing Paris-green, in the proportion of one or two teaspoonfuls to a pailful of water, would destroy many of them; but the safest way is to catch and kill the enemy. Where a plant is seen suddenly to wilt and die, the author of the mischief can generally be found within a few inches of the plant destroyed, and a short distance below the surface of the ground. These larvæ are all vigorously attacked by various species of parasites.

No. 196.—The Spotted Paria.

Paria sex-notata (Say).

This is a small beetle, about three-tenths of an inch long, pale in color,—sometimes dark,—having the wing-covers spotted with black, and ornamented with regular rows of dots, which disappear towards the tip (see Fig. 343); beneath it is

Fig. 343.

blackish. It is a stout insect, with a polished surface, and is very active in its movements, hopping briskly about when approached or disturbed. The beetle appears at the time when the fruit is partly grown, which, in the northern parts of the continent, is towards the end of May. When these insects are abundant, they devour the leaves of the plants with such avidity that they are soon completely riddled with holes, and the crop of fruit materially injured.

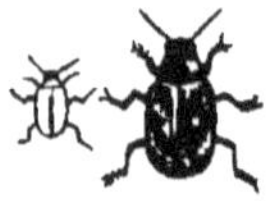

Remedies.—On account of the advanced growth of the fruit when the beetle appears, it would be unsafe to use strong poisons, such as Paris-green. It would be much safer to use hellebore, and quite effectual; probably air-slaked lime, soot, or ashes dusted on the foliage would also remedy the evil.

No. 197.—The Striped Flea-beetle.

Phyllotreta vittata (Fabr.).

This pretty little beetle, although most commonly found on

Fig. 344.

young turnips and cabbages, is sometimes found also eating the leaves of strawberry-plants. The beetle, which is shown magnified in Fig. 344, is less than one-tenth of an inch long, black, with a broad, wavy, yellowish stripe on each wing-cover. It is very active, leaping away to a considerable distance when an attempt is made to catch it.

The larva, which is also shown in the figure, is found on the roots of young cabbage-plants; it is about one-third of an inch long, white, with a dusky line on the anterior half of its body. The head is pale brown, and on the posterior extremity is a brown spot equal to the head in size. When the larva reaches maturity, it forms a little earthen cocoon near its feeding-place, and in this transforms to a pupa (Fig. 344) of a whitish color, from which, in a few days, the beetle appears.

The remedies recommended for the spotted Paria, No. 196, are equally applicable in this case.

No. 198.—The Canadian Osmia.

Osmia Canadensis Cresson.

This is a small four-winged insect which occasionally proves destructive to strawberry-plants. In Fig. 345 it is shown much magnified; its natural size is indicated by the short line at the side of the figure. The head, thorax, and abdomen in both sexes are green, and more or less densely covered with short hairs, those on the thorax being longest. The wings are nearly transparent, with blackish veins. The female is larger than the male.

FIG. 345.

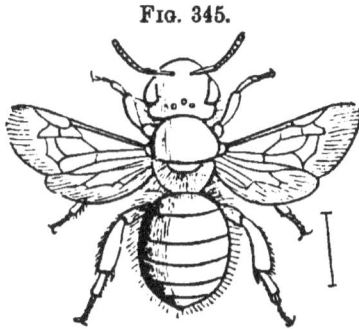

These insects nibble away the leaves, chewing the fragments into a sort of pulp, and carrying it away to be used in the construction of their nests. The injury done to strawberry-plants by them is sometimes very marked.

No. 199.—The Strawberry Leaf-stem Gall.

This is an elongated gall, an inch or more in length, found on the stalk of the leaf of the strawberry near its base, produced by an undetermined species of gall-fly. Its surface is irregular and its color red, while the internal structure is spongy. If these galls are opened about the middle of July, there will be found in each, about the centre, a small, milk-white, footless grub, semi-transparent, with a smooth, glossy skin, a wrinkled surface, and a few fine, short hairs. Its jaws are pale brown, and its length at this period is about one-sixteenth of an inch, the body tapering a little towards each extremity. This insect doubtless changes to a pupa within the gall, from which flies escape later in the season, or early the following spring.

No. 200.—The Strawberry Saw-fly.

Emphytus maculatus Norton.

This insect in the perfect state is also a four-winged fly, which in the larval condition is very destructive to the leaves of the strawberry. The accompanying figure, 346, illustrates the insect in its various stages; 1 shows the under side of the pupa, 2 a side view of the same, 3 the perfect fly, all

Fig. 346.

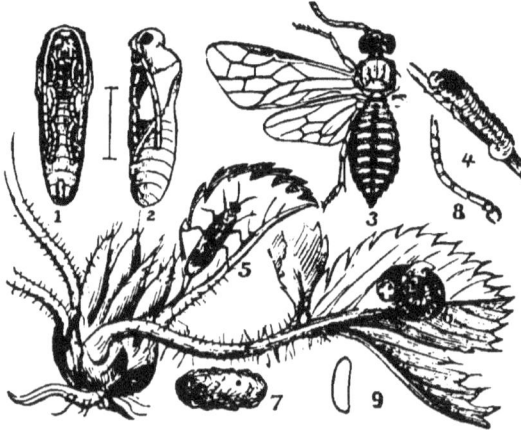

magnified; 4 the larva crawling, 6 the same at rest, 5 the perfect insect with its wings closed, and 7 the cocoon, all of the natural size; 8 one of the antennæ, and 9 an egg, both magnified. The egg is placed within the substance of the stem of the leaf early in May by means of the peculiar saw-like apparatus with which the female is provided. It is about one-thirtieth of an inch long, and of a white color; its presence produces a slight swelling on the stalk, and by splitting the stalk so as to open the swelling the egg may be found. The eggs absorb moisture from the stem and increase in size, and in about a fortnight hatch, when the young larvæ at once begin to feed on the leaves. At first they attract but little attention, as the holes they make in the leaves are small, but as they increase in size they often completely riddle the foliage and destroy its usefulness.

When full grown, they are nearly three-fourths of an inch long, of a pale-greenish color, with a faint whitish bloom. The skin is semi-transparent, revealing the movement of the internal organs, which show through as dark-greenish patches. There is a broken band along each side, of a deeper shade of green, and below this the body has a yellowish tint. The head is yellowish brown, with six black dots, the jaws dark brown, and the under surface yellowish. The larvæ fall to the ground when disturbed.

When mature, they burrow under the surface, and form oval cocoons by cementing together minute fragments of earth, and within these enclosures the remaining transformations are completed, the insect finally issuing in the perfect or winged form.

The fly is black, with two rows of large whitish spots upon the abdomen; antennæ black, legs brown. The wings, when spread, measure a little more than half an inch across. Those belonging to the first brood of larvæ appear on the wing early in July, when eggs are deposited for a second brood, which are found during August. They complete their larval growth, enter the ground, and construct their earthen cells, in which they remain unchanged until the following spring, when they enter the pupa state and transform to flies within a few days.

Remedies.—Hellebore and water, or Paris-green and water, showered on the vines in the proportions recommended under No. 181, will destroy them.

ATTACKING THE FRUIT.

No. 201.—The Stalk-borer.

Gortyna nitela Guenee.

This larva, which is commonly found in the stalks of the potato and tomato, may be said to have a rather varied taste, as it also bores into the stalks of the dahlia, aster, and cockleburr, the cob of the Indian corn, and the fruit of the strawberry. In Fig. 347 we have a representation of the larva.

When it leaves the fruit or other substance it has occupied, it descends a little below the surface of the earth, and in a few days changes to a brown chrysalis, from which the moth (Fig. 348) emerges from about the end of August to the middle of September.

In case this insect should so multiply as to require a remedy, hand-picking is the only one suggested.

SUPPLEMENTARY LIST OF INJURIOUS INSECTS WHICH AFFECT THE STRAWBERRY.

ATTACKING THE ROOTS.

The larva of the goldsmith beetle, No. 77, and also that of the May beetle, No. 113, attack the roots of the strawberry. The latter, which is commonly known as the white grub, is frequently very destructive.

ATTACKING THE LEAVES.

The oblique-banded leaf-roller, No. 35; the climbing cut-worms, No. 45; the tarnished plant-bug, No. 71; the horned span-worm, No. 86; the grape-vine Colaspis, No. 153; and the currant Angerona, No. 210.

ATTACKING THE FRUIT.

The flea-like negro-bug, No. 185, is not uncommon on the fruit of the strawberry.

INSECTS INJURIOUS TO THE RED AND WHITE CURRANT.

ATTACKING THE STEMS.

No. 202.—The Imported Currant-borer.

Ægeria tipuliformis Linn.

This insect has for many years been a serious impediment in the way of successful currant-culture. It is an importation from Europe, where it has long proved troublesome; in the larval state it burrows up and down the interior of the stems, making them so hollow and weak that they frequently break in the spring from the weight of foliage when swayed by the action of the wind.

The parent of this destructive larva is a pretty, wasp-like moth (see Fig. 349), which measures, when its wings are expanded, about three-quarters of an inch across.

FIG. 349.

The body is of a bluish-black color, the abdomen being crossed by three narrow golden bands, while on the thorax and at the base of the wings are streaks of a similar color. The wings are transparent, but veined and bordered with brownish black with a coppery lustre; the bordering is widest on the front wings, which are also crossed by a band of the same color beyond the middle. The moth appears about the middle of June, when it may be found in the hot sunshine, darting about with a rapid flight, sipping the nectar of flowers or basking on the leaves, alternately expanding and closing its fan-like tail, or searching for suitable places in which to deposit its eggs.

The female is said to lay her eggs near the buds, where in a few days they hatch into small larvæ, which eat their way to

336

the centre of the stem, where they burrow up and down, feeding on the pith all through the summer, enlarging the channel as they grow older, until at last they have formed a hollow several inches in length. When full grown, the larva (*b*, Fig. 350) is whitish and fleshy, of a cylindrical form, with brown head and legs, and a dark line along the middle of its back. Before changing to a chrysalis, a passage is eaten nearly through the stem, leaving merely the thin outer skin unbroken, thus preparing the way for the escape of the moth.

FIG. 350.

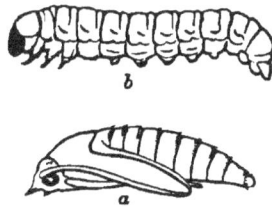

Within this cavity the larva changes to a chrysalis (*a*, Fig. 350, where both larva and chrysalis are shown magnified). Early in June the chrysalis wriggles itself forward, and, pushing against the thin skin covering its place of retreat, ruptures it, and then partly thrusts itself out of the opening, when in a short time the moth bursts its prison-house and escapes, soon depositing eggs, from which larvæ are hatched, which carry on the work of destruction.

While this insect chiefly infests the red and white currant, it attacks the black currant also, and occasionally the gooseberry. Where the hollow stems do not break off, indications of the presence of the borers may be found in the sickly look of the leaves and the inferior size of the fruit.

Remedies.—In the autumn or spring all stems found hollow should be cut out and burnt. During the period when the moths are on the wing they may often be captured and destroyed in the cool of the morning, at which time they are comparatively sluggish.

No. 203.—The American Currant-borer.
Psenocerus supernotatus (Say).

This borer is the larva of a beetle, and, although belonging to an entirely different order from No. 202, is very similar in its habits, but it may be distinguished by its smaller size and by the absence of feet. It is a small, white, cylindrical, footless larva, with a brown head and black jaws, which also feeds upon the pith of the stems, rendering them hollow and often killing them. Usually several, and sometimes as many as eight or ten, of these borers are found within the same cane. The change to a pupa takes place within the stalk, and in the latter part of May or early in June the perfect insect escapes.

This is a small, narrow, cylindrical, brownish beetle. (See Fig. 351, where it is represented magnified, the outline figure at the side showing the natural size.) The wing-cases are of a darker brown behind the middle; there is a whitish dot on the anterior part of each elytron, and a large, slightly oblique, and sometimes crescent-shaped spot of the same color just behind the middle; the antennæ are slender, and nearly as long as the body. The beetle flies during the day, but is much less active than No. 202, and hence more easily captured. The cutting out and burning of the infested stalks will be found of great advantage in this instance also. This borer is sometimes attacked by parasites.

FIG. 351.

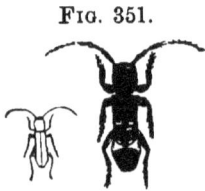

No. 204.—The Currant Bark-louse.

Lecanium ribis Fitch.

Early in the spring there are sometimes seen on the bark of currant-stems brownish-yellow, hemispherical scales, about one-third of an inch in diameter, under which will be found a quantity of minute eggs: as the season advances, these hatch, when the young lice distribute themselves in all directions over the twigs, puncturing them with their beaks, and absorbing the sap.

Another species, called the Circular Bark-louse, *Aspidiotus circularis* Fitch, is mentioned by Dr. Fitch as occurring on currant-stalks in the form of minute, circular, flat scales, about one-thirtieth of an inch in diameter.

These lice may be removed by scraping the stems or applying to them a strong alkaline wash.

ATTACKING THE LEAVES.

No. 205.—The Imported Currant-worm.

Nematus ventricosus Klug.

This is the larva of one of the saw-flies, and is perhaps the most troublesome of all the insects the currant-grower has to encounter. It is a European insect, first noticed in America in 1858, and within the comparatively brief period which has since elapsed it has spread over a large portion of the continent. This insect usually passes the winter in the pupal condition, but occasionally in the larval state.

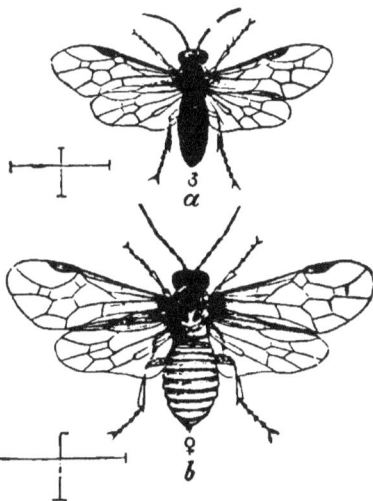

Fig. 352.

Very early in the spring the flies appear. The two sexes differ materially in appearance. In Fig. 352, *a* represents the male, and *b* the female, both enlarged, the lines at the side indicating their natural size. The male approaches the common house-fly in size, but the body is scarcely so robust, and the wings, four in number, are more glossy. Its body is black, with a few dull-yellow spots above, the under side of the abdomen being yellowish and the legs bright yellow ; the veins of the wings are black or brownish black. The female is larger than the male, and differs in the color of its body, being mostly yellow instead of black. These flies are active only during the warmer parts of the day; at other times they are quiet or almost torpid.

Within a few days the female deposits her eggs on the under side of the leaves on the larger veins in rows, as shown in Fig. 353. When first laid, they are about one-thirtieth of an inch long, but they either absorb moisture from the leaf, or else the expansion is due to the development of the enclosed larva, and within four or five days they increase in length to about one-twentieth of an inch, are rounded at each end, whitish and glossy. In about ten days the young larva hatches, and it is then about one-twelfth of an inch long, of a whitish color, with a large head, having a dark, round spot on each side of it. At first they eat small holes in the leaves, as shown at 2 and 3 in the figure, feeding in companies of from twenty to forty on a leaf, so that soon the leaf is completely destroyed, all its soft parts being consumed, and nothing but the skeleton frame-work remaining. Shortly they increase in size, and, parting company, spread in all directions over the bush, first changing to a green color, then to green with many black dots, and finally to plain green again, tinged with yellow at the extremities, just before the change to the pupa takes place. When from half to two-thirds grown, they are extremely voracious, and will, when numerous, often strip an entire bush of its leaves in the course

FIG. 353.

of two or three days. They are represented at this stage of their growth in Fig. 354. When mature, they are about three-quarters of an inch long, at which time they seek for a suitable spot in which to form their cocoons.

Fig. 354.

These are sometimes made among dry leaves or rubbish on the surface of the ground, sometimes under the ground, and occasionally attached to the stems or leaves of the bush on which they have fed. The location once fixed on, the larva begins to contract in length, and spins a cocoon over itself, which, when finished, is nearly oval, smooth, of a brownish color and papery texture, within which it changes to a small, delicate, whitish-green pupa, very transparent, with the encased limbs and wings of the future fly distinctly visible, from which the fly escapes late in June or early in July. Soon again eggs are deposited, from which another brood of larvæ are sent forth on their destructive mission, completing their growth before summer closes, and in most instances changing to pupæ before winter.

The flies composing the separate broods do not all appear at once; some are weeks later than others, keeping up a regular succession, and making continual watchfulness necessary in order to save the foliage from destruction. They feed on the cultivated gooseberry as readily as on the currant, and also on the wild varieties of gooseberry.

Remedies.—A minute parasitic fly has been found attacking the eggs by Prof. Lintner, of Albany, N. Y., closely resembling, if not identical with, the insect represented in Fig. 181.

The presence of this parasite may be detected by the discoloration of the eggs, which become brown. A species of Ichneumon, *Hemiteles nemativorus* Walsh, is parasitic on the larva, while the placid soldier-bug, *Podisus placidus* Uhler, also destroys the larva.   This friendly insect, which is

FIG. 355.

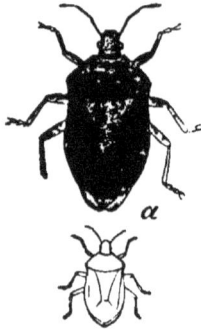

shown magnified at *a* in Fig. 355, and of the natural size in the outline below, has the head, thorax, and legs black, and the abdomen red, with an elongated black spot in the centre, crossed by a whitish line. It approaches a larva, thrusts its proboscis into its victim, and sucks it until it shrivels and dies.   An average-sized bug will consume several of these larvæ every day, and, where they are plentiful, must prove a material check to the increase of the saw-fly.   The aphis lions, the larvæ of the gauze-wing flies, *Chrysopa* (see Fig. 132, under No. 57), also devour them.

Notwithstanding these various aids among insects, it is usually necessary to employ other remedial measures, and nothing is more efficient than powdered hellebore mixed with water, in the proportion of an ounce to a pailful, and sprinkled freely on the bushes.   If thoroughly applied, most of the larvæ will be found dead or dying within an hour afterwards. If hellebore is not at hand, hot water may be used, a little hotter than one can bear the hand in, showered plentifully on the bushes.   This will not injure the foliage, but will dislodge most of the larvæ, and when on the ground they can be trodden on and destroyed.   Hand-picking may also be resorted to, especially while the insects are young and feeding in groups of twenty to forty on a leaf.   An experienced eye will soon detect them, usually on the lower leaves of the bushes, the little holes in the leaves aiding in their discovery.

No. 206.—The Native Currant Saw-fly.

Pristiphora grossulariæ Walsh.

Although this is not a very common insect, it has been reported as destructive from several localities.   In its perfect state it is also a saw-fly, resembling the imported species (see *b*, Fig. 356), yet there are differences which the entomologist can readily detect, that place this insect in a different genus; such as the arrangement of the veins on the wings, the close resemblance of the sexes, and the marked difference in the relative size of

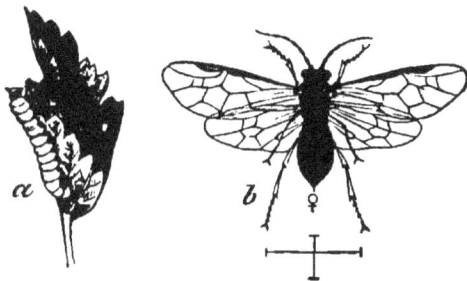

FIG. 356.

the two insects, the native species being but two-thirds the size of the imported one in all its various stages.

The larva (*a*, Fig. 356) of this species is always green, and is never ornamented with black spots, which are so numerous on the imported insect as it approaches maturity; neither do the young larvæ gather in large numbers on one particular leaf, but are from the first scattered over the bushes.   There are two broods in the year; the first one may be looked for about the end of June, and the second during the latter part of August.

The cocoons, which are similar in appearance to those of the imported saw-fly, but smaller, are usually constructed among the twigs and leaves of the bush on which the larvæ have fed.

The winged insects, of which the female is represented in the figure, have the body black, with yellow markings; the second brood are said to come out of the pupa the same season, which, if correct, involves the conclusion that the eggs are laid on the stems of the currant-bushes late in the autumn.

Where these insects prove troublesome, they may be subdued with the same remedies as are recommended for No. 205.

No. 207.—The Ohio Currant Saw-fly.

Pristiphora rufipes St. Fargeau.

This insect is referred to in Dr. Fitch's twelfth "Annual Report" as entomologist for New York State, as occurring in the vicinity of Cleveland, Ohio, in 1858. The larvæ are of a pea-green color, with black heads; they live together in clusters, and eat the leaves, beginning at the edge and devouring all except the coarser veins. As they move they spin a very light web from leaf to leaf, and they are said to let themselves down to the ground, when disturbed, by a fine thread of silk. When mature, they are three-eighths of an inch long, the segments of the body are slightly wrinkled, and along each side is a row of protuberances or warts of the same color as the body. When ready for their next change, they enter the ground and form small oval cocoons, within which they change to pupæ.

The fly is black, with transparent wings and light-brown legs.

No. 208.—The Currant Span-worm.

Eufitchia ribearia (Fitch).

In many districts this is a very common insect; it may be easily distinguished from the saw-fly caterpillars by its peculiar mode of progression, arching its body into a loop at every step; in Fig. 357 the larva is represented in various attitudes. When disturbed, it lowers itself suddenly by a silken thread from the bush on which it has been feeding, and remains suspended in mid-air until the threatened danger is past, when it regains its former position. It is a native insect, and is frequently found on the wild currant and gooseberry bushes in the woods. When full grown, the caterpillar

measures an inch or more in length, is of a whitish color, with
a wide yellow stripe down the back, another of the same character along each side,
and a number of black
spots of different sizes
upon each segment.
The under side is
white with a slight
tinge of pink, is also
spotted with black,
and has a wide yellow
stripe down the mid-
dle. There is but one
brood of this insect in
a year; hence there is
no probability of its
ever becoming so for-
midable a pest as the
imported saw-fly.

The eggs, which are
very pretty (see Fig.

Fig. 357.

358, which shows óne much magnified at *a*, and others of
the natural size at *b*), are attached to the stems and twigs in
the autumn, and remain in this condition
until spring, when they hatch about the
time the bushes are in full leaf, the larvæ
attaining their full growth within three
or four weeks. When ready for their
next change, they descend to the ground,
and, having penetrated a short distance
under the surface, change to dark-brown
chrysalids about half an inch long (see 3,

Fig. 358.

Fig. 357), in which condition they remain two or three weeks,
or more, when the perfect insects are liberated.

The moth (Fig. 359) is of a pale-yellowish color, with
several dusky spots, which vary in size and form, being more

prominent in some specimens than in others, forming sometimes one or two irregular bands across the wings.   When

Fig. 359.

expanded, the wings measure about
an inch and a quarter across.   Within
a brief period the female deposits her
eggs for the next year's brood on the
twigs and branches, where they endure the heat of the remaining portion of the summer without hatching,
and the piercing cold of the succeed-
ing winter without injury, awaiting the arrival of their proper
time for development the following spring.

Remedies.—Powdered hellebore, which is so speedy and
certain a remedy in the case of the saw-flies, does not act with
the same promptitude in this instance.   This larva seems to
be much hardier and more difficult to destroy with poisonous
substances ; hence, if hellebore is used, the liquid should be
made twice or three times the usual strength.   Paris-green is
more certain and effectual where there is no objection to its
use.   Hand-picking is more practicable with this larva, on
account of its habit of letting itself down by a strong silken
thread and remaining suspended ; and if after striking the
bush a forked stick is passed all around under it, all the
hanging threads may be caught, and the larvæ drawn out in
groups and crushed with the foot.   This insect is quite destructive to the black currant, and also to the gooseberry.

No. 209.—The Spinous Currant Caterpillar.

Grapta progne (Cram.).

The parent of this caterpillar is a very handsome butterfly, which is shown in Fig. 360 ; the pair of wings
which are attached to the body show the upper surface, the
detached pair the under surface.   Above, the fore wings are
of a dull reddish orange, widely bordered on the outer edge
with dark brown, while within there are many spots of brown
and black.   The hind wings are dark brown, tinged with red

behind, shading into reddish towards the front. The under side of both wings is dark brown, traversed by many grayish lines and streaks, and on the anterior pair there is a very wide band towards the outer edge of a paler color. The wings are very irregular in outline, with many notches and prominences; when expanded, they measure an inch and a half or more across.

FIG. 360.

This butterfly passes the winter in the perfect or winged state, hiding in some sheltered nook, where it remains torpid during the winter, awakening to life again with the genial warmth of spring. It may be found very early in the season skipping about with a peculiar jerky flight around the openings in woods, occasionally resting on the sunny side of a tree, or stopping to sip the sweet juice exuding from the stump of a freshly-cut tree.

The eggs are laid on currant and gooseberry bushes, both wild and cultivated, and when hatched the larvæ do not feed in groups, but singly on the leaves. When full grown, they are about an inch and a quarter long, and vary in color from a light brown to a dull greenish yellow, with narrow black and yellow lines. The body is thickly covered with long branching spines, which also vary in hue, some being yellow, others orange, and some dark brown, many of their branches being tipped with black.

When full grown, the larva seeks some secluded spot in which to change to a chrysalis; sometimes the under side of a leaf or twig is selected, and there, after spinning on the surface a small web of silk, its hind legs are hooked in the fibres, and it remains suspended head downwards. The body soon contracts in length, and in two or three days the caterpillar skin is shed, and a rugged, angular-looking chrysalis appears, of a brown color prettily ornamented with silvery spots.   After remaining in the pupal condition from one to two weeks, the time varying with the heat of the weather, the butterfly appears.

There are two broods during the season, the larvæ of the first one appearing late in June, those of the second maturing early enough in the autumn to admit of the escape of the butterfly before severe frost occurs.   This insect rarely appears in sufficient numbers to prove troublesome; should it become numerous, hellebore and water would no doubt prove an efficient remedy, or the larvæ might be subdued by hand-picking.

No. 210.—The Currant Angerona.

Angerona crocataria (Fabr.).

The moth from which this caterpillar is produced is usually quite common, but the larva, although often found feeding on currant leaves, feeds upon the gooseberry, strawberry, and other plants besides, and hence is seldom sufficiently abundant on

Fig. 361.

currant-bushes to attract much attention.   The accompanying figure, 361, represents the larva a little more than two-thirds grown, feeding on a gooseberry leaf.   At this period it does not differ materially from the full-grown larva except in size.

When mature, it is about an inch and a half long or more, tapering towards the front.   It is of a yellowish-green color, with an indistinct whitish line down the back, and a rather broad whitish streak on each side below the spiracles, bordered above with faint purple, which increases in depth of color on the hinder segments and becomes a purple stripe on the last one.   The spiracles are white, edged with purple; each segment of the body has its anterior portion swollen and yellowish, and on most of the segments there are a few minute black dots.

When the larva has attained its full size, it draws together the edges of a leaf half-way or more, and, forming a slight net-work of silken threads, changes to a chrysalis of a dark olive-green color, with a pale-greenish abdomen, a row of black dots down the back, and another on each side, from which in about ten days or a fortnight the perfect insect appears.

The moth (Fig. 362) is a native of America; it flies by day, and may often be seen on the wing about openings in the borders of the forest. Its wings are yellow, varying in shade from deep to pale, with dusky spots and dots sometimes few in number, while in other specimens they are very numerous, the larger ones being so arranged as to form an imperfect band across the wings. The under side is usually a little deeper in color than the upper, and, when the wings are expanded, they measure nearly an inch and a half across.

FIG. 362.

In its native haunts the larva probably feeds on the wild currant, gooseberry, and strawberry. Although a common insect, this is rarely complained of as injurious; should it become so, the remedies recommended for No. 181 would no doubt be found efficient.

No. 211.—The Currant Amphidasys.

Amphidasys cognataria Guenee.

The larva of this insect is also a measuring-worm or looper, and, although seldom found in sufficient numbers to prove destructive, instances are on record where currant-bushes have been almost stripped of their leaves by them. The larva, when full grown, is about two inches long, and may, when not feeding, usually be found clinging to one of the leaves or branches by its hind legs, with its body extended straight out, so that it might easily be mistaken for the stem of a leaf. Its body is pale green, with a darker, interrupted green line down the back, indistinct, broken transverse lines of the same color, and a yellow cross line on the posterior end of each segment. There are two small tubercles on the segment immediately behind the head, and the body is dotted with very small whitish tubercles and a few short black hairs. In some specimens there is a small brown tubercle on each side behind the middle, and a purplish-brown ridge on the last segment.

When mature, the larva descends to the ground and buries itself in the earth, where it eventually changes to a chrysalis

Fig. 363.

about seven-tenths of an inch long and of a dark-brown color, from which the moth escapes the following spring.

This is a handsome moth (see Fig. 363), which, when its wings are spread, will measure two inches or more across. Both fore and hind wings are gray, dotted and streaked with black, and with a wavy light band crossing the wings beyond the middle. The under surface is paler than the upper; the body gray, dotted with black.

This insect is a very general feeder, and on that account is not likely ever to prove very destructive to the currant; it has been found feeding also on the plum, Missouri currant, red spirea, and maple.

No. 212.—The Four-striped Plant-bug.

Poecilocapsus lineatus (Fabr.).

This is a bright-yellow bug, about three-tenths of an inch long, with black antennæ and two black stripes on each of its wing-covers, the outer one on each side terminating in a black dot. In Fig. 364 this insect is represented magnified, with an outline the natural size. It punctures the young leaves of the currant-bushes on both their upper and under surfaces, causing small brown spots, not much larger than pin-heads, but these are sometimes so numerous and closely placed that the leaves become completely withered. The insects are very active, and when approached drop quickly to the ground or fly away. They begin to feed in May or June, and continue for a month or two, often disfiguring the bushes very much and retarding their growth. When very troublesome, they may be captured by visiting the bushes early in the morning, and, while torpid with cold, brushing them off into a pail partly filled with water on which a little coal-oil has been poured. They do not confine their attacks to currant-bushes, but often injure the dahlia by puncturing the flower-stems and causing them to wither; they also affect the weigelia, the deutzia, and other shrubs.

FIG. 364.

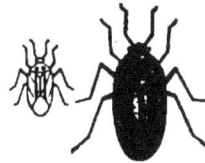

No. 213.—The Currant Plant-louse.

Aphis ribis Linn.

Towards midsummer there often appear on the leaves of red-currant bushes blister-like elevations of a brownish-red color, while on their under sides are corresponding hollows, in which will be found a multitude of lice, some of a pale-yellowish color, without wings, others with transparent wings, and bodies marked with black.

In the position these insects occupy they are very difficult to destroy, except by hand-picking the leaves and burning them. A few lady-birds, such as are referred to under No. 57, introduced among them, will speedily lessen their numbers. These lice rarely inflict any serious injury, but for a time give the bushes an unsightly and diseased appearance: they are an importation from Europe, where they have long been injurious to the currant.

ATTACKING THE FRUIT.

No. 214.—The Currant Fruit-worm.
Eupithecia interrupto-fasciata Packard.

This insect is readily distinguished from the gooseberry fruit-worm by the number of its legs, which are only ten, while the gooseberry fruit-worm has sixteen. The currant fruit-worm is a span-worm; that is, it arches its body, when in motion, with every step. When full grown, it is about five-eighths of an inch long, and varies in its color and markings. Its body is pale greenish-ash, or yellowish green, with a dark-colored line down the back, and another on each side, but occasionally this latter is wanting. Sometimes there is a row of dark-colored, lozenge-shaped spots along the dorsal line, and in some instances there is a second lateral line lower down the side. On the hinder part of the terminal segment there are two short greenish spines. The head varies in color from yellowish or greenish to light brown; the under side of the body is white or pale greenish, with a yellow line in the middle.

When full grown, it draws several leaves or other suitable protecting material together, fastens them with silken threads, and within the enclosure changes to a chrysalis, from which eventually the moth escapes.

The fore wings of the moth are of a bluish-gray color, with a bluish dot near the centre of each, and a dark line crossing them immediately beyond the dot.

No. 215.—The Currant Fly.
Epochra Canadensis (Loew).

This insect is occasionally found attacking the fruit of both the red and the white currant. In its perfect state it is a small two-winged fly, which lays its eggs on the currants while they are small; the larva enters them while still green, and feeds on their contents, leaving a round, black scar at the point of entry. The affected currants ripen prematurely, and shortly decay and drop to the ground, when, on opening them, there will be found in each a small white grub, about one-third of an inch long, which, when mature, leaves the currant and probably passes the pupa state under the ground.

SUPPLEMENTARY LIST OF INJURIOUS INSECTS WHICH AFFECT THE RED AND WHITE CURRANT.

ATTACKING THE BRANCHES.

The oyster-shell bark-louse, No. 16, so common on the apple, is sometimes said to be destructive to currant-bushes.

ATTACKING THE LEAVES.

The fall web-worm, No. 27; the Cecropia emperor-moth, No. 28; the oblique-banded leaf-roller, No. 35; the saddle-back caterpillar, No. 49; the Io emperor-moth, No. 112; the yellow woolly-bear, No. 146; and the currant Endropia, No. 216, are all found feeding on currant leaves.

ATTACKING THE FRUIT.

The gooseberry fruit-worm, No. 219.

23

ATTACKING THE LEAVES.

No. 216.—The Currant Endropia.

Endropia armataria (Herr. Sch.).

About the middle of July there will sometimes be found on black-currant bushes small, nearly black, geometric caterpillars, dotted and marked with pale yellow, and with a series of crescent-shaped whitish spots down the back, and a row of raised dark-brown dots along each side, those on the hinder segments tipped with yellow, while on the last segment there is a fleshy hump or prominence composed of two round tubercles. When full grown, this larva is about three-quarters of an inch long, when it constructs a slight web, interweaving portions of dead leaves or other rubbish, and within this changes to a brown chrysalis, in which condition it remains throughout the winter, producing the perfect insect the following June.

The moth is represented in Fig. 365, about the natural size. Its wings are yellowish brown shaded with purple, especially on the hind wings, and with streaks and dots of a deeper shade of brown. The under surface is deep yellow dotted and streaked with reddish brown.

Fig. 365.

This insect is by no means common, and hence is never likely to prove generally injurious to currant-bushes. Although it prefers the black currant, it feeds also on the leaves of the red currant.

354

No. 217.—The Red Spider.

Tetranychus telarius (Linn.).

This is a very small mite, which often proves a serious pest to gardeners, especially to those who cultivate plants under glass. Occasionally, in dry weather, it attacks the leaves of the black currant and destroys them. Fig. 366 represents the male of this species, very much enlarged, the mite itself being scarcely visible to the unaided eye; the small dot within the circle at the side of the figure indicates the natural size of the insect. It spins a web on the under side of the leaves, of threads so slender as to be scarcely visible even with an ordinary magnifying-glass until woven into a net-work. Under this shelter will be found a colony, consisting of mature individuals of both sexes and young mites of all ages. By the aid of their jaws,

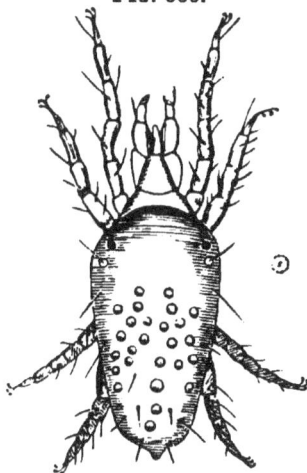

FIG. 366.

which are not unlike the beak of a bird, they tear away the surface of the leaf, and plunge their beaks into the wound and suck the juice.

The egg of this mite is nearly round, and colorless; the larva is a minute, transparent object, not unlike its parent, but it has only six legs, and creeps along slowly. The mature mites have eight legs, and vary much in color, some being greenish marked with brown specks, others rust-colored or reddish, and many of them brick-red.

The leaves attacked soon indicate the presence of this invader by their sickly hue; the sap being sucked by a multitude of tiny mouths, they soon assume a yellowish cast, with patches of a grayish or lighter shade; and if the mite is allowed to pursue its course unchecked, the foliage becomes much injured, and sometimes is destroyed. It is said to pass the winter under stones, concealing itself there when the leaves on which it has fed have fallen.

Remedies.—Various preparations of sulphur and soap have been recommended, used separately or together, mixed with water, and applied to the bushes with a syringe. Plain soap and water, or water alone, freely applied, is regarded by some as efficient, as the insect is known to thrive best in a dry atmosphere. In applying any liquid, it is necessary to wet the under side of the leaves in order to make the application effectual, since if applied to the upper surface only the mites would remain uninjured beneath.

SUPPLEMENTARY LIST OF INJURIOUS INSECTS WHICH AFFECT THE BLACK CURRANT.

ATTACKING THE STEMS.

The imported currant-borer, No. 202.

ATTACKING THE LEAVES.

The fall web-worm No. 27 ; and the currant span-worm, No. 208.

INSECTS INJURIOUS TO THE GOOSEBERRY.

ATTACKING THE BRANCHES.

No. 218.—The Mealy Flata.

Pœciloptera pruinosa Say.

This is a small, four-winged bug, which attacks the succulent shoots of the gooseberry, and sometimes the leaves,
sucking the juices. It is wedge-shaped, about one-third of an
inch long, almost twice as high as wide, of a dusky bluish
color, covered with white, meal-like powder, its
wing-covers showing some faint white dots, and Fɪɢ. 367.
near their base three or four dusky ones.

The insect is shown in Fig. 367; it is not confined to the gooseberry, but is found on the grape, also on the
privet and on various other shrubs.

ATTACKING THE FRUIT.

No. 219.—The Gooseberry Fruit-worm.

Dakruma convolutella (Hübn.).

This injurious insect spends the winter in the chrysalis state,
enclosed in a snug, brown, papery-looking cocoon, shown at *a*
in Fig. 368, which is hidden among leaves or other rubbish on
the surface of the ground. During the
latter part of April the moth appears.
(See *b*, Fig. 368.) Its wings, when
expanded, measure nearly an inch
across. The fore wings are pale gray,
with dark streaks and bands; there
is a transverse diffuse band a short distance from the base of
the wing, enclosing an irregular whitish line, which terminates

Fɪɢ. 368.

before it reaches the front edge of the wing. Near the outer edge is another transverse band, enclosing a whitish zigzag line; there is also a row of blackish dots within the outer margin, while the veins and their branches are white; the hind wings are paler and dusky. The head, antennæ, body, and legs are all pale gray, whiter below than above.

The insect deposits its eggs probably on the young gooseberries shortly after they are set. The egg soon hatches, when the young larva burrows into the berry, where it remains safely lodged; as it increases in size it fastens several of the berries together with silken threads, sometimes biting the stems off some of the berries, so that they may be more readily brought into the desired position, and within this retreat revels on their substance at its leisure. The larva makes but one hole in a berry, and that barely large enough to admit its body. When disturbed, it displays great activity, and works its way backwards out of the fruit very quickly, and drops part way or entirely to the ground by a silken thread, by means of which, when danger is past, it is enabled to recover its former position. It is shown, suspended and on the fruit, in Fig. 369. When fully grown, this intruder is

FIG. 369.

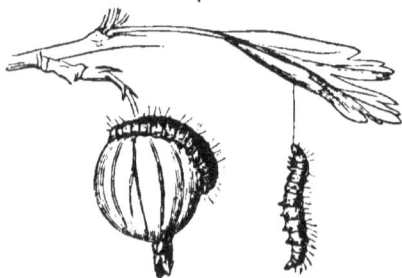

about three-quarters of an inch long, the body thickest in the middle, tapering slightly towards each extremity. It is of a pale-green color, sometimes with a yellowish or reddish tint, glossy and semi-transparent. The head is small, pale brown, and horny-looking, and on the upper surface of the next segment is a patch of the same color and appearance.

When ready for its next change, which is usually before the fruit ripens, it lowers itself to the ground, and there spins its little silken cocoon among leaves or rubbish, as already stated, and remains as a small, brown chrysalis within the cocoon until the following spring. There is only one brood of these insects during the year.

The infested fruit soon indicates the presence of the larva by becoming discolored, and, if sufficiently grown, it ripens prematurely, otherwise it becomes of a dull whitish color, and soon withers. This pest also attacks the wild gooseberry, as well as the currant, both the white and the red variety. In this latter case, since the fruit is not large enough to contain the body of the larva, it draws the clusters together, and, fastening the berries to each other with silken threads, lives within the enclosure.

Remedies.—The most satisfactory method of destroying this insect is by hand-picking, and its habits are such that its presence is easily detected. Any berries found coloring prematurely should be carefully examined, and, as the larvæ slip out and fall to the ground very quickly, watchfulness is needed to prevent their escape in this manner. Where neglected, they often increase to an alarming extent, and in some instances half the crop or more has been destroyed by them. It is recommended to let chickens run among the bushes after the fruit has been gathered, so that they may devour the chrysalids; any leaves or rubbish under the bushes should also be gathered and burnt, and a little lime or ashes scattered over the ground in their place. Dusting the bushes freely with air-slaked lime early in the spring, and renewing it if washed off by rain, will also in great measure deter the moths from depositing their eggs on the young fruit then forming.

No. 220.—The Gooseberry Midge.

Cecidomyia grossulariæ Fitch.

This second enemy to the fruit is a very small, two-winged fly, which punctures the young gooseberry and deposits its tiny eggs therein. These eggs develop into minute, bright-yellow larvæ of an oblong-oval form, much resembling the midge which is found in the ear of wheat. The larva changes to a pupa within the fruit, and the perfect fly escapes during the latter part of July.

The fly is scarcely one-tenth of an inch long, measuring from the head to the tips of its closed wings; it is of a pale-yellow color, with black eyes, blackish antennæ, and transparent wings tinged with dusky brown.

It is probable that those flies which come out during the latter part of July deposit eggs for a second brood in some later fruit or other suitable substance, and that the larvæ mature, change to pupæ, and pass the winter under ground, producing flies the following spring.

Remedies.—All fruit found prematurely decaying or assuming an appearance of ripeness before the time of ripening should be gathered and burnt, with all fallen gooseberries. By careful attention to this matter both of the insects which injure the fruit may be kept in subjection.

SUPPLEMENTARY LIST OF INJURIOUS INSECTS WHICH AFFECT THE GOOSEBERRY.

ATTACKING THE LEAVES.

The imported currant-worm, No. 205; the currant span-worm, No. 208; and the spinous currant caterpillar, No. 209, all feed on the leaves of the gooseberry as freely as they do on those of the currant.

INSECTS INJURIOUS TO THE MELON.

ATTACKING THE ROOTS.

No. 221.—The Squash-vine Borer.

Ægeria cucurbitæ Harris.

This borer is the larva of a moth belonging to the group known as Egerians, or Clear-wings, which have the greater portion of their wings transparent, and hence closely resemble wasps. They are active in the daytime, and enjoy the warmth of the summer's sun.

The moth, which is represented in Fig. 370, is a very pretty object. Its body is about half an inch long, orange-colored or tawny, with four or five black spots down the back ; the fore wings are olive-brown and opaque, the hind wings transparent, except the margins and veins ; the hind

FIG. 370.

legs are densely fringed with long reddish and black hairs, and the wings, when expanded, measure an inch or more across.

This active enemy deposits her eggs on the stems of the young vines near the roots about the time they begin to run, or soon after, where the young larva, when hatched, bores into the stem and devours the interior. The full-grown larva (Fig. 371) is about an inch long, tapering towards each extremity, soft, of a whitish color, and semi-transparent, with a dark line down the back, caused by the internal organs showing through the transparent

FIG. 371.

skin ; there are a few short hairs on each segment, arising singly from small, hard, warty points. The head is small, of a brown color, and there is a patch of a similar shade on the next segment.

When full grown, the larva leaves the plant and seeks shelter under the earth, where it forms an oblong-oval cocoon (Fig. 372) of particles of earth fastened together with gummy silk, within which it transforms to a shining, brown chrysalis, which remains unchanged until the following season. When the perfect insect is about to escape, the chrysalis wriggles itself part way out of the cocoon, so that the moth when freed from the chrysalis shell may find no further obstacle to its exit.

Fig. 372:

The presence of this borer in the vines is soon manifested by a sickly appearance and a drooping of the foliage, which, if the cause is not removed, soon results in withering and death. Whenever a vine becomes unhealthy, the stems should be examined, and cut into if necessary, to remove the lurking enemy. The moths may be prevented from depositing their eggs by lightly banking up the young vines with earth, as they grow, as far as the first blossoms. When once the larva is within the stem, no other remedy than the knife is of much service.

ATTACKING THE STEMS.

No. 222.—The Striped Squash Beetle.

Diabrotica vittata (Fabr.).

This is a troublesome enemy to the melon-grower, and is destructive not only to the melon, but also to the squash and cucumber, boring in the caterpillar state into the lower part of the stem, and sometimes down into the root, while the perfect beetle feeds on the tender leaves of the young plants, and injures the buds and young shoots of later growth.

Fig. 373.

The parent beetle, shown in Fig. 373, magnified, makes its appearance very early in the season, as soon as the young seed-leaves of the vines are above ground, and sometimes even penetrates the earth a little in search of the sprouting seeds. The female lays her eggs probably on the stem of the vine, just above or below the surface, and from the egg is soon hatched a young larva, which eats its way to the centre of the stem and consumes its substance. When full grown, it is about four-tenths of an inch long, slender, but little thicker than an ordinary pin, of a whitish color, with a small, brownish head, and the end of the body suddenly truncated. Fig. 374 shows this larva highly magnified ; *a* a back view, *b* a side view. The first brood of the larvæ mature in June and July, or in about a month after the eggs are laid ; they then leave the vines and penetrate into the earth, where each one forms a little cavity for itself, in which it changes to a pupa. Both back and front views of the pupa are given in Fig. 375, magnified. It is about one-fifth of

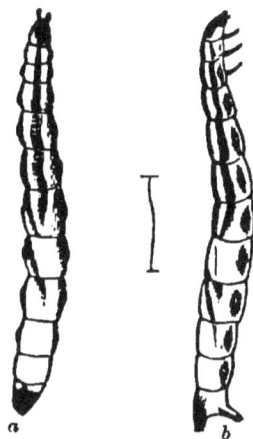

Fig. 874.

a   *b*

an inch long, of a whitish color, with two spines at the extremity of the abdomen. After remaining in the pupal state about a fortnight, the perfect insect escapes, and works its way out of the cell and up to the surface of the ground.

The beetle is about a quarter of an inch long, of a bright-yellow color, with a black head, and broad stripes of black on the wing-covers, which are also punctated with rows of dots. The feet and the under side of the abdomen are black. There are two or three broods during the year, and the larva has been found in the stems of the melon-vines as late as October.

Fig. 375.

The winter is passed in the ground in the pupal condition. The beetles may often be found in considerable numbers in the autumn in the flowers of melon, squash, and pumpkin vines, feeding on the pollen and other portions of the flower. They have also been known to attack the blossoms of the pear and cherry.

Remedies.—The best remedy is to prevent the access of the beetle by covering the young vines with small boxes, open at the bottom and covered at the top with muslin. Sprinkling the vines with a mixture of Paris-green and flour, in the proportion of one part of the former to twenty parts of the latter, air slaked lime, plaster of Paris, soot, and ashes, have all been recommended and used with more or less advantage. The larvæ should also be searched for and destroyed; the time to look for the first brood is when the vine is beginning to run. If the stem close to the root, and the root itself, are found smooth and white, the plant is uninjured; but if they are roughened or corrugated on the surface, and of a rusty color, the presence of the insect is indicated.

A parasitic two-winged fly, a species of Tachina, attacks the beetles, depositing its eggs on their bodies, from which hatch small fleshy larvæ, which eat their way into the abdomen of their victims and eventually destroy them.

ATTACKING THE LEAVES.

No. 223.—The Cucumber Flea-beetle.

Crepidodera cucumeris (Harris).

Although a very small insect, this is not to be despised. It is a beetle, about one-sixteenth of an inch long, with a black body, finely punctated, and clothed with a whitish pubescence; there is a deep transverse furrow across the hinder part of the thorax; the antennæ are of a dull-yellow color, and the legs of the same hue, except the hinder pair of thighs, which are brown; these latter are very thick and strong, and well adapted for leaping. Fig. 376 represents this insect much magnified; the short line at the side indicates its natural size. The beetles pass the winter concealed under stones or rubbish, appear very early in the season, and attack the young melon and cucumber plants as soon as they are up. They eat small round patches on the upper surface of the leaves, consuming their substance, but not always eating entirely through. They hop very actively from leaf to leaf, and are very destructive to young plants; while partial to melon and cucumber vines, they are also fond of the potato, raspberry, turnip, cabbage, and other plants.

Fig. 37

Their larvæ are minute and slender, tapering towards each end, and are said to live within the substance of the leaves attacked; hence the plants suffer from the depredations of the larvæ as well as from the injuries caused by the beetles. They attain maturity, pass through the pupa state, and change to beetles, within a few weeks, and there is a constant succession of the insect in its various stages throughout the greater part of the summer.

Remedies.—Air-slaked lime, powdered hellebore, or Paris-green mixed with flour, in the proportion of one part of the poison to twenty or thirty parts of flour, dusted on the foliage, will speedily destroy them.

No. 224.—The Melon Caterpillar.

Eudioptis hyalinata (Linn.).

This is an insect which is very widely distributed, being found throughout the greater part of North and South America. In some parts of the Southern States it is particularly destructive. The larvæ, which are shown feeding on the leaves in Fig. 377, are, when mature, about an inch and a quarter long, translucent, and of a yellowish-green color, with a few scattered hairs over their bodies. They are not content to feed on the leaves only, but eat into melons, cucumbers, and pumpkins at all stages of growth, sometimes excavating shallow cavities, and at other times penetrating directly into the substance of the fruit. They spin their cocoons in a fold of the leaf of the melon, as shown in the figure, or on any other plant growing near by, and change to slender, brown chrysalids, about three-quarters of an inch long, from which, in a short time, the perfect insect is produced.

The moth, which is also represented in Fig. 377, is very beautiful. The wings are of a pearly-white color, with a

Fig. 377.

peculiar iridescence, bordered with black, and they measure, when expanded, about an inch across. The body and legs are of the same glistening white, and the abdomen terminates in a movable brush-like tuft of a pretty buff color, tipped with white and black. The number of broods of this insect during the year has not been definitely ascertained ; the winter is passed in the chrysalis state.

Remedies.—If the first brood of young worms occur before the melons have attained half their growth, powdered helle-bore mixed with water, in the proportion of an ounce to two gallons of water, and sprinkled on the vines, may be safely used to destroy them. Strong tobacco-water would also prob-ably have the same effect, while on small patches they could doubtless be killed by hand. Two species of parasitic insects are known to prey on them : one is a species of Tachina fly, the other an Ichneumon fly, *Cryptus conquisitor.* (See Fig. 42, where it is referred to as a destroyer of the apple-tree tent-caterpillar, No. 20.)

ATTACKING THE FRUIT.

No. 225.—The Neat Cucumber Moth.
Eudioptis nitidalis (Cram.).

Another common name for this insect is the "pickle-worm," which has been given to it in consequence of its larva being often found in pickled cucumbers. This larva is about an inch long, trans-lucent, and of a yel-lowish-white color tinged with green; on each segment

FIG. 378.

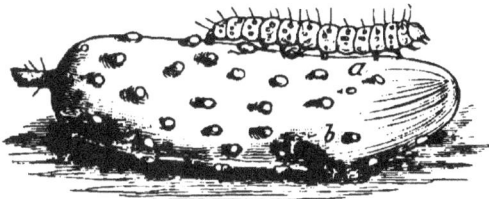

there are a few slightly-elevated shining dots, from each of which issues a fine hair; the head is yellow, margined with brown. Fig. 378 represents this larva, with a young cucum-ber into the side of which it has bored. These caterpillars are very destructive in some of the Western States. They begin to appear about the middle of July, and continue their destructive work until late in September; they attack the fruit, boring cylindrical holes in it, and feed on the flesh.

Sometimes three or four larvæ will be found in the same fruit, while the presence of a single specimen will often cause the cucumber to rot.

When mature, the larva leaves the fruit, and, drawing together a few fragments of leaves on the ground, spins a slight cocoon, within which it changes to a slender, brown chrysalis, from which the moth issues in eight or ten days. The insects forming the late brood pass the winter in the chrysalis state. The moth (Fig. 379) is of a yellowish-brown color, with a

FIG. 379.

purplish reflection, the fore wings having an irregular patch, and the hind wings the greater portion of their inner surface yellow. The under side has a pearly shade; the thighs, breast, and abdomen below are silvery white; the other portions of the legs are yellow. The body of the female terminates in a small, flattened, black brush, squarely trimmed, the segment preceding it being of a rusty-brown color above. The male has a much larger brush-like appendage, formed of long, narrow scales, some of which are whitish, some orange, others brown.

Remedies.—This insect is a difficult one to control. If the vines are carefully watched about the time the early brood appear, the larvæ may be destroyed by hand while still small; but if not discovered until after they have penetrated the fruit, the infested melons or cucumbers should be gathered and fed to hogs or scalded.

No. 226.—The 12-Spotted Diabrotica.

Diabrotica 12-punctata (Oliv.).

FIG. 380.

This beetle also is occasionally destructive to melons and squashes, eating into their substance. It is a yellow beetle, with twelve black spots, represented in Fig. 380. It is closely related to the striped squash beetle, No. 222.

ATTACKING THE LEAVES.

No. 227.—The Cranberry Worm.

Rhopobota vacciniana (Packard).

This larva is very injurious to the foliage of cranberry-vines, and, on account of the devastation it causes, has received in some localities the significant name of the "fire-worm." It hatches in the Eastern States from the 20th of May to the 1st of June, from eggs which have remained upon the vine all winter. These are found on the under side of the leaves in masses having the form of a flat circular scale of a pale-yellow color.

FIG. 381.

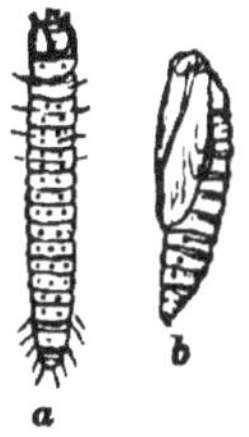

The larva, which is shown at *a*, Fig. 381, is green, with a few fine hairs scattered over the surface of its body. It feeds upon the tender growing shoots, drawing the leaves together, fastening them with silken threads, and concealing itself within the enclosure. When full grown, it spins a slight cocoon, either among the leaves on the vines or amidst leaves and rubbish on the ground, and there changes to a chrysalis, as shown at *b* in the figure. The pupa state lasts from ten to twelve days.

FIG. 382

The moth (see Fig. 382) is of a dark ash-color, the fore wings whitish, dusted with brown and reddish scales, with narrow white bands on the front edge, alternating with broader yellowish-brown bands, five of which are larger than the others, and from four of these, distinct but irregular lines cross the wings. The tips of the fore wings are dark brown and pointed.

24 369

The hind wings are dusky gray. The moths are very numerous during the month of June, when eggs for a second brood are deposited, the larvæ from which appear early in July, succeeded by the perfect insect, which deposits the eggs that remain dormant until the following spring.

Remedies.—For all cranberry insects flooding is the most effectual remedy; the vines should be kept under water for two or three days, which will clear them for the time entirely from all insect pests. Where this is not practicable, the vines may be showered with a mixture of Paris-green and water, in the proportion of a teaspoonful of the poison to two gallons of water. Fires also may be lighted to attract and destroy the moths.

No. 228.—The Glistening Cranberry Moth.

Teras oxycoccana (Packard).

This moth, the larva of which is said to feed on cranberry-vines, measures, when its wings are spread, nearly three-fourths of an inch across. Its fore wings are of a uniform reddish-brown color, with a peculiar shining appearance, the red tint being due to scattered bright-red scales; there are no other spots or markings. The hind wings are glistening gray. The body is of a dark slate-color, with a pale tuft of hairs at the tip of the abdomen. This insect is said to be merely a variety of No. 36.

No. 229.—The Yellow Cranberry Worm.

Teras vacciniivorana (Packard).

In the cranberry-fields of New Jersey this is a common insect. The larva, which is shown magnified in Fig. 383, both back and side views, draws the leaves together, fastens them with silken threads, and feeds upon their upper surface. It is of a pale-yellow color, with a slight greenish tinge, and a few fine, long, pale hairs arising from prominent tubercles. When mature, it is nearly three-tenths of an inch long. The caterpillar changes to a brown chrysalis within the leafy en-

closure, which, when the moth is about to escape, protrudes
partly out of its hiding-place. The pupa is about a quarter

FIG. 383.

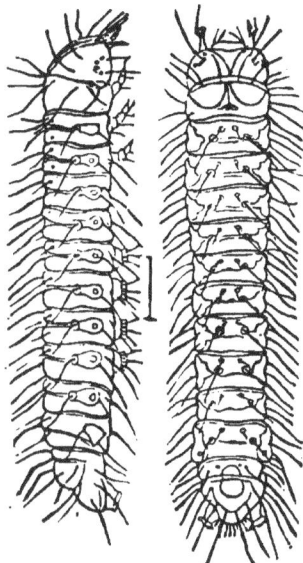

FIG. 384.

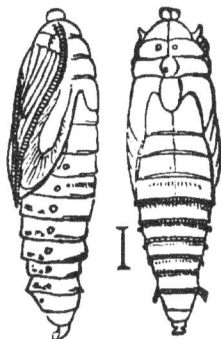

of an inch long, and is represented from two different as-
pects in Fig. 384, both much
magnified.

The moth measures, when its
wings are spread, about half
an inch across; both front and
hind wings are yellow, mottled with a deeper ochreous shade.
This also is said to be a variety of No. 36.

For remedies, see No. 227.

No. 230.—The Red-striped Cranberry Worm.

This larva, which is shown in Fig. 385, has been observed
by Dr. Packard injuring the heads of cranberry-plants in
Massachusetts. It draws and fastens the leaves together and
feeds on their upper surface, and sometimes constructs a tube
of silk between two leaves, when the latter are severed from
their connection with the branch and held in place by silken
threads. In these instances the leaves speedily wither and
turn brown, and it often happens that the tips of vines over
large patches will present a brown and withered aspect from
this cause.

The larva (see Fig. 385) is less than half an inch long, slender, and tapering a little towards each extremity, of a pale-

Fig. 385.

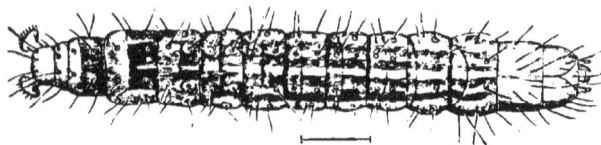

green color, with six longitudinal pale-reddish lines, which are broken and irregular on the anterior segments, and more distinct and wider on the hinder part of the body. On each segment there are several small black tubercles, from each of which arises a single hair. The moth is undescribed.

For remedies, see No. 227.

No. 231.—The Cranberry Span-worm.

Cidaria Sp.

In Massachusetts, and especially in the vicinity of Harwich, this larva has proved very injurious, having in one instance entirely stripped the foliage of about two acres of cranberry-vines. It very much resembles the larva of the canker-worm, and is about the same size; its color is dull reddish brown, with longitudinal lines and many dots of dark brown. There is a broad dusky band just above the spiracles; the under side is paler than the upper. When full grown, it measures about eight-tenths of an inch in length. The moth has not been described.

For remedies, see No. 227.

No. 232.—The Hairy Cranberry Caterpillar.

Arctia Sp.

This is a caterpillar which sometimes injures cranberry-vines in New England. It is about an inch and a half long, is covered with yellowish-gray hair, and has longer tufts of darker hair at each end of the body. It devours the leaves of the young growing shoots, often depriving them entirely of foliage.

No. 233.—The Cranberry Saw-fly.

Pristiphora identidem Norton.

This insect, which is closely allied to the imported currant-worm, No. 205, is destructive to cranberry-vines on Cape Cod. The perfect insect is a saw-fly, the female having a toothed ovipositor, with which she makes a slit in the leaves, depositing an egg therein. Broods of the larvæ appear early in June, and again in August. When first hatched, they are pale yellowish green, but become darker with age; the head is black in the young specimens, lighter in the full-grown ones. When mature, they measure about three-tenths of an inch long, are cylindrical and smooth, with two lighter, whitish-green stripes running the whole length of the body. Towards the end of June they spin their cocoons among withered leaves or other rubbish, from which flies are produced about ten days afterwards.

The perfect insect has the body black, the legs marked with yellowish red and black, the wings transparent, with black veins.

No. 234.—The Cranberry Gall-fly.

Cecidomyia Sp.

About the middle of June the small leaves at the tips of the growing shoots may often be found fastened together. Within these clusters is a small, pinkish or orange-colored larva, having the form shown at *b* in Fig. 386, which is without legs, and when first hatched is white. This larva spins a cocoon (see *a* in the figure), which resembles white tissue-paper; this is formed among the small leaves at the end of the shoot, and within it the insect changes to a pupa, as shown at *c*.

In about twelve days the perfect insect, a gall-gnat, appears (see *d*, Fig. 386; *e* represents the antenna of the female, much enlarged). This gnat is found in almost every cranberry-bog. There are not usually more than two of these larvæ on any one shoot, and often there is only one. The mischief done consists mainly in the killing of the extreme tip of the vine, which prevents the formation of a fruit-bud for the next year's growth, unless, as is sometimes the case, the vine by an extra effort puts them out at the side.

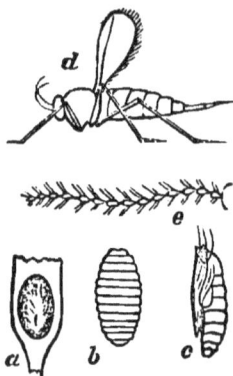

FIG. 386.

Remedies.—There is a little Chalcis fly parasitic on this insect, which destroys it in large numbers. The measures recommended under No. 227 will also be applicable here.

No. 235.—The Cranberry Aphis.

There is a large, red plant-louse which sometimes occurs on cranberry-vines and punctures the leaves and tender stems, to their manifest injury. This aphis is destroyed by the larva of a small lady-bird, a species of Scymnus, which larva is oval in form, and covered with a white fuzz on its back. Flooding will destroy this aphis also.

No. 236.—The Cranberry Spittle Insect.

Clastoptera proteus Fitch.

This is a small, soft insect, with legs, but without wings, which is found in the early part of June in little masses of froth upon growing shoots of the cranberry-vine. The froth is the sap of the plant sucked in and then exuded by the young larva, probably for concealment. The insect belongs to the order *Homoptera,* having no jaws, but a beak, through which it sucks the sap of the plant.

The perfect insect jumps with the agility of a flea, and is found hopping about among the vines. It seldom occurs in sufficient numbers to inflict material injury. It is found also on the blueberry.

ATTACKING THE FRUIT.

No. 237.—The Cranberry Fruit-worm.

This is the caterpillar of a small moth related to the leaf-rollers, and is shown in Fig. 387. It is of a yellowish-green color, and appears early in August, when it injures the fruit, entering berry after berry, eating the inside of each, and making it turn prematurely red. It attains its full growth by the beginning of September, when it buries itself in the ground, where it forms a cocoon covered with grains of sand, scarcely to be distinguished from a small lump of earth, within which it changes to a chrysalis. Flooding is the only remedy suggested for this insect.

Fig. 387.

No. 238.—The Cranberry Weevil.
Anthonomus suturalis Lec.

About the middle of July, or just before the blossoms are ready to expand, this weevil appears. It is a small, reddish-brown beetle, with a dark-brown head and a beak half as long as its body, shown in Fig. 388. The thorax is a little darker than the wing-covers, and is sparingly covered with short whitish hairs; the wing-cases are ornamented with rows of indented dots. The beetle is a little over one-eighth of an inch long, including the beak. Having selected a blossom-bud about to expand, it drills a hole through the centre with its snout, in which is deposited a pale-yellow egg. The bud is then cut off by the beetle at the stem, and drops to the ground, and within it the egg hatches to a dull-white grub with a yellow head and black jaws (see Fig. 388), which feeds upon the bud, and, passing through its transformations, produces the perfect beetle, which eats its way out, leaving a round hole in the side of the de-

Fig. 388.

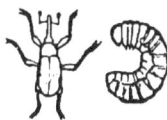

caying bud to mark its place of exit. The beetles sometimes, though seldom, feed upon the berries. They may be destroyed by flooding with water. There is a minute Chalcis fly which is parasitic on the larvæ and destroys numbers of them.

Since many of the insects most injurious to the orange attack alike the branches, the leaves, and the fruit of the tree, and sometimes the trunk also, the grouping of the species, carried out when treating of the enemies of other fruits, will not be attempted with those of the orange. The insects belonging to each order will be brought together and treated consecutively, beginning with the Lepidoptera, which includes butterflies and moths. The remedies for scale-insects, as they apply alike to all the different species, will be referred to towards the end of this section.

No. 239.—The Cresphontes Butterfly.

Papilio cresphontes Fabr.

In the perfect state, this is a large and handsome butterfly, which measures, when its wings are spread, from four to five inches across. The wings are black above, with an irregular, triangular band of broad yellow spots, covering a considerable portion of their surface, as shown in Fig. 389. The hind wings have two long, projecting points or tails, with an oval yellow spot on each; they are also notched, and have the indentations marked with yellow. The under side is yellowish, with dusky veins and markings, and a row of crescent-shaped blue spots on the hind wings. The body is black above, yellow at the sides and beneath.

The eggs are globular, and are deposited singly on the leaves. The young caterpillars are very much like the full-grown ones in form and color, but the gray markings are darker, and the white blotches not so large as in the mature larva. When full grown, it is about two and a half inches long, and very peculiarly marked. (See Fig. 390.) Above

877

it is dull brown, almost covered with irregular whitish
blotches spotted with brown.   The first four segments have

Fig. 389.

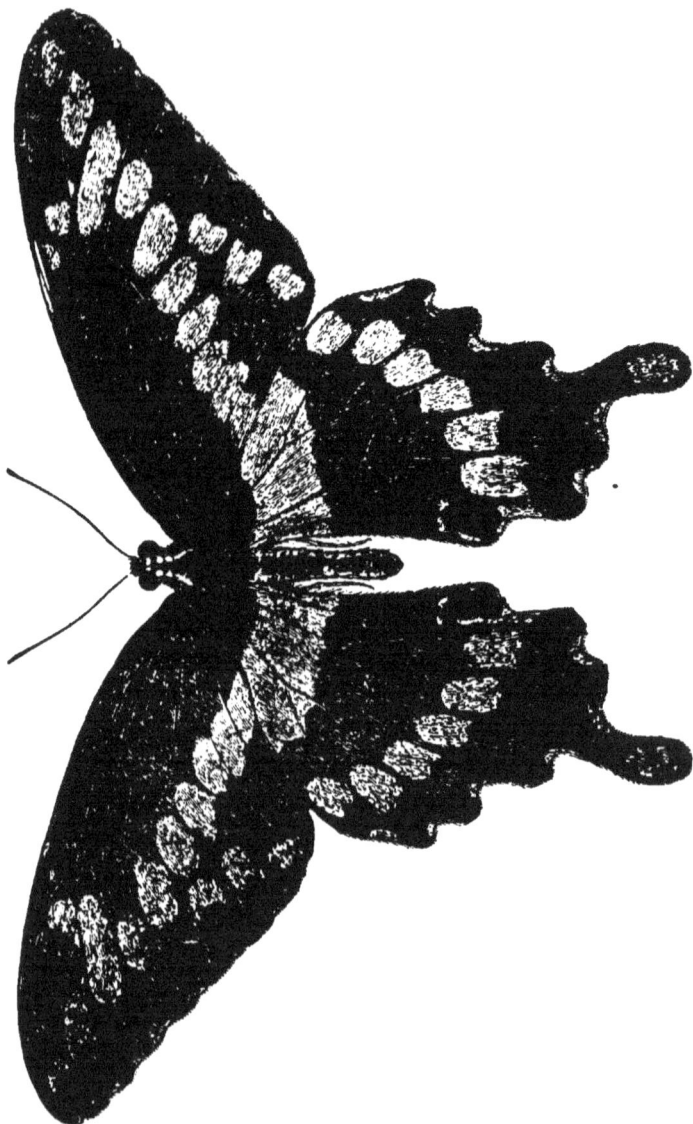

on each side a longitudinal white band; from the fourth to
the eighth is a large white patch, nearly oval in form, more

or less dotted with brown ; another similar white or cream-colored patch, with brownish dots, covers the posterior por-

Fig. 390.

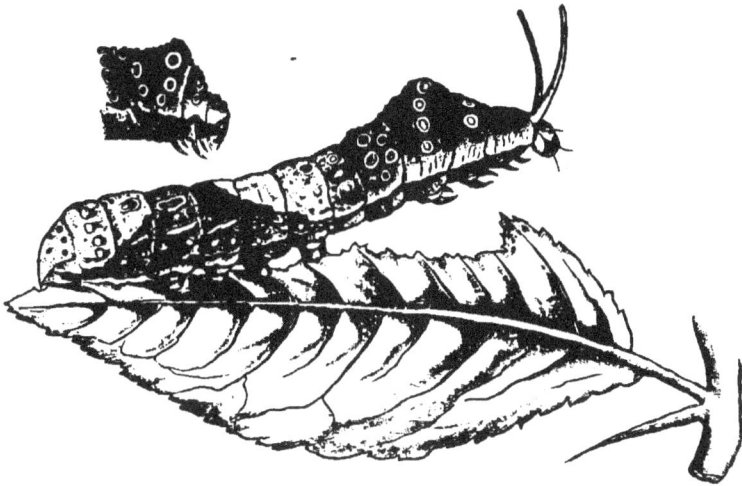

tion of the body. Behind the head there are two long, red, fleshy horns, which can be protruded at will, and these, when extended, emit a very disagreeable odor, which probably serves to protect the caterpillar from its enemies. The under side of the body is of a brownish color. The larva completes its growth in about a month, when it changes to a chrysalis. This is nearly an inch and a half long (see Fig. 391), irregularly forked at its upper end, with a prominent point upon its breast, and a loop of silk around the middle ; the hinder extremity is also fastened to the supporting twig or branch, hooked in a tuft of silk. Its color is gray and brown, of

Fig. 391.

varying shades, and so exactly resembles that of the bark of the orange-tree that it is extremely difficult to detect. In from eight to sixteen days after the chrysalis is formed the butterfly emerges.

In Florida there are usually four broods of the butterflies in the course of the summer, the last brood wintering in the chrysalis state, from which the butterflies emerge in April. The caterpillar, which is commonly known as "the orange dog" in Florida, devours the foliage of orange-trees, sometimes seriously injuring young trees by stripping them bare. It may easily be subdued by hand-picking, as its large size and singular appearance promptly lead to its discovery.

Within the past ten years this butterfly has extended its range very much, and it is now comparatively common throughout the Northern and Western States, and in the warmer parts of Canada. In the North it feeds chiefly on prickly ash, *Zanthoxylum Americanum.*

No. 240.—The Orange Basket-worm.

Platœceticus Gloveri Packard.

During the month of February this insect is found upon the orange-trees in different parts of Florida. The larva

Fig. 392.

forms an oblong-oval case of a paper-like substance, interwoven with bits of leaves or bark, as shown in Fig. 392; within this it lives. When full grown, it is a little over half an inch long, thick and fleshy, and varies in color from light brown to a much darker shade. The head is marked with dark and light wavy lines, and is protruded from the case, along with the anterior segments, when the larva· is feeding or moving from place to place. The case of the female is about one-fourth larger than that of the male. Both of these are shown in the figure.

On reaching maturity, the case is suspended from a leaf or twig, and within it the larva changes to a dark-brown chrysalis ; the chrysalis of the male works its way partly out of the case at the lower end, where, after the escape of the moth, the empty pupa-skin remains.

The male moth (Fig. 392) is dark brown, sometimes nearly black, with delicate wings, small body, and feathered antennæ, and measures, when its wings are spread, about six-tenths of an inch across. The female is wingless, of a whitish color, and transforms within the case, where, also, the eggs are laid, the young larvæ, when hatched, escaping from the orifice at the lower end. This insect has also been found feeding on the leaves of the fig.

The conspicuous cases constructed by the larvæ are easily seen, when they may be picked and destroyed.

No. 241.—The Orange Leaf-roller.

Platynota rostrana (Walker).

During the growing season the edges of the young leaves of orange-trees are often found rolled up into a sort of tube. These tubes are formed by a small, yellowish-green caterpillar, which, when full grown, is about three-quarters of an inch long, with a brown head, and a polished plate of the same color on the next segment, a dark stripe down the back, and an indistinct dark line along each side. It is active in its movements, lives within the tube it constructs, and feeds upon the foliage.

The larva changes to a brown chrysalis, nearly half an inch long, within the case, from which in a few days a moth escapes.

The male differs from the female in the markings on its fore wings. All the wings of both sexes have a ground-color resembling that of cork, but the fore wings of the male have a dark-brown stripe along the front edge, expanding into a large spot of the same color towards the tip of the wing, while the fore wings of the female have minute dark-brown tufts, arranged in lines more or less distinct, running obliquely across them. The wings of the male measure, when spread, nearly three-quarters of an inch across; those of the female are a little larger. This leaf-roller has been found troublesome in several localities in Florida. Where it exists in such abundance as to require a remedy, hand-picking should be resorted to, or the trees should be syringed with powdered hellebore and water, or Paris-green and water, as recommended under No. 181.

No. 242.—The Orange-leaf Nothris.

Nothris citrifoliella Chambers.

In the larval form this is a cylindrical yellow caterpillar, with a black head, and a black patch on the next segment. It feeds upon the half-grown leaves of the new shoots of the orange, fastening them together with silken threads. It also frequently devours the terminal buds, and thus materially injures the growth of the tree. When full grown, it is about half an inch long, very quick in its movements, and if disturbed lets itself down from the twig by a silken thread, by means of which it is enabled to regain its former position among the leaves when danger is past.

When ready for its next change, the larva rolls up a portion of a leaf, and spins within the enclosure a delicate silken cocoon, in which it changes to a dark-brown chrysalis. The moth is found late in August and early in September; it is of a grayish ochreous color, the fore wings streaked with reddish and dotted with brown, the hind wings pale gray with a reddish tint. The body is ochreous, dotted with dull red.

Should this insect at any time become so abundant as to require the use of remedies, those suggested for No. 241 will be applicable.

No. 243.—The Orange Leaf-notcher.

Artipus floridanus Horn.

This is a beetle which is represented magnified in Fig. 393, the line below it indicating the natural size.   It eats jagged notches in the leaves of the orange, as shown in the figure, disfiguring and injuring the foliage.   It is about a quarter of an inch long, of a pale greenish-blue or copper color, and densely clothed with white scales.   The thorax is unevenly dotted, and there are on the wing-cases ten longitudinal lines of dots of varying sizes, divided by slight ridges.   The under side of the body and legs is also scaly and hairy.

Fig. 393.

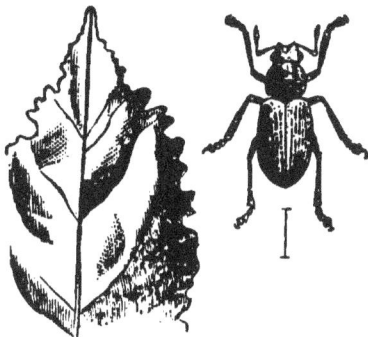

In some localities in Florida these beetles are said to be very abundant.   As they readily drop when the trees are jarred, they may be easily collected on sheets spread under the trees.

No. 244.—The Angular-winged Katydid.

Microcentrum retinervis Burm.

There is, perhaps, no insect of large size so destructive to the foliage of the orange as this.   It is a large green katydid, and one of the commonest insects in the South.

During the daytime it is seldom seen, as it is then hidden among the thick foliage of trees and shrubs, but towards dusk it leaves its hiding-places and makes the air resonant with its music, which is produced by rubbing the wings against the thighs.   The eggs are deposited in abundance upon both twigs and leaves, as shown in Fig. 394 at 1 *a* and 2 *b*, overlapping each other.   They are of a long, oval form,

FIG. 394.

and nearly flat. The young katydids issue from that end of the egg which projects beyond the leaf, leaving the empty egg-shell still in position behind. When first hatched, they feed only upon the surface of the leaf, but as they increase in size they devour the whole substance. When mature, they acquire wings, which enable them to fly readily from tree to tree, appearing as shown at 1 in the figure. From the head to the extremity of the closed wings, the full-grown insect measures about two and a half inches. The outer wings are green, with leaf-like veinings, the under pair of a paler green, and beautifully netted; the antennæ are long and thread-like, and the hind legs slender. The female is furnished with a curved ovipositor at the end of the abdomen.

Fortunately, there is a small Chalcid fly parasitic on the eggs of this katydid, which, when mature, is little more than one-eighth of an inch long; it is the *Eupelmus mirabilis* of Walsh. The female, which is shown at 2, Fig. 394, has dusky wings, and an abdomen which she can elevate over her thorax in a peculiar manner. The male is represented at 2 *a* in the same figure. The eggs of this parasite are placed within the eggs of the katydid, where the larvæ hatch and undergo their transformations, issuing as flies from circular holes which they cut through the egg-shells, as shown at 2 *b*. A large proportion of the eggs of the katydid are parasitized by this insect.

Remedies.—Collect the eggs during the winter and place them in boxes covered with coarse wire gauze until spring, so that the parasites may be permitted to escape. Several species of birds are said to devour these katydids.

No. 245.—The Lubber Grasshopper.

Romalea microptera Serv.

This is a large species of locust, very destructive to orange-leaves, which has received the common name of "the lubber grasshopper" from its sluggish habits. When full grown, it is about two and a quarter inches long, of a yellow color, the

25

wing-cases shaded with rosy pink and barred and spotted with black. The larvæ are shaped like the mature insects, but have no wings. They are black, and are striped and banded with orange-yellow. The wings of the perfect insect (see Fig. 395) are so short—reaching only half-way to the

FIG. 895.

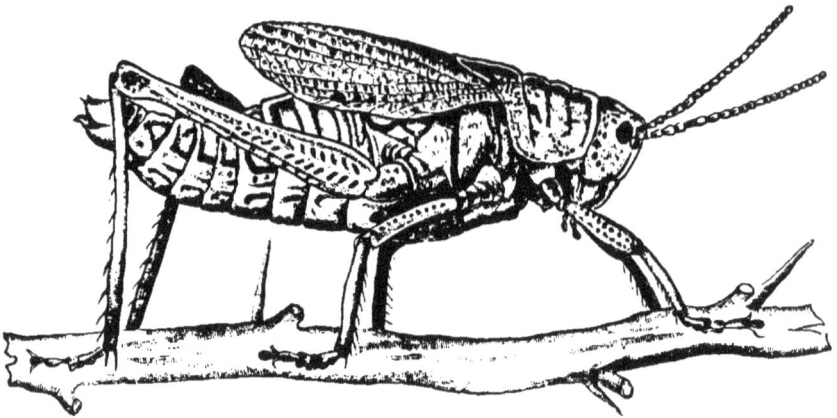

extremity of the abdomen—that they are quite useless for the purpose of flight. Their eggs are deposited in the ground. Since they cannot fly, they may easily be destroyed by hand.

No. 246.—The Leaf-footed Plant-bug.

Leptoglossus phyllopus (Linn.).

FIG. 896.

The leaf-footed plant-bug is of a reddish-brown color, with a long, sharp beak, and a transverse yellowish-white band across its wing-covers. The wings, when raised, show the body, which is of a bright-red color, with black spots. The shanks of the hind legs are flattened out into leaf-like append-ages, as shown in Fig. 396. This insect is said to puncture the tender shoots and ter-minal branches of the orange-tree, often killing them. It also injures ripe plums, by puncturing them and sucking portions of their contents. Notwithstanding its injurious habits, it has been by some writers classed among beneficial insects as a destroyer of the harlequin cabbage-bug.

No. 247.—The Cotton-stainer.

Dysdercus suturellus H. Schf.

This insect, like that last described, belongs to the order of true bugs (*Hemiptera*); it is commonly known as the red-bug, or cotton-stainer, and is one of the worst pests with which the cotton-planters of Florida and the West Indies have to contend. It injures the cotton by piercing the stems and bolls and sucking the sap; but the principal injury to the crop is occasioned by its staining the cotton in the opening bolls with its excrement. It also attacks the fruit of the orange, puncturing the rind, sucking the juice, and causing the fruit to decay and fall to the ground. When full grown, it is from six to seven tenths of an inch long, and appears as shown in Fig. 397, the thorax triangular, with its anterior part red, posterior portion black, all margined with whitish yellow. The scutellum is triangular, red, margined with pale yellow; the wing-cases are flat, with two distinct whitish lines crossing them, which intersect each other near the centre; they are also partly margined with a yellowish line. The under side is bright red, with yellowish-white markings on the edge of each segment.

FIG. 897.

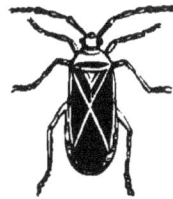

Each female produces about one hundred oval, amber-colored eggs, which are attached in clusters to the under side of the leaves. The young bugs are bright red, with black legs and antennæ. These bugs are usually found in immense numbers, and where cotton has been planted between the rows of orange-trees instances are recorded where a large proportion of the oranges have been destroyed. The mature insects often gather in great numbers on heaps of cotton-seed, when they may be killed by pouring scalding water upon them.

No. 248.—The Orange Aphis.

Siphonophora citrifolii Ashmead.

In Florida this species of plant-louse is very prevalent, and is found during the spring and summer months in various stages of development, clustering on the tender shoots and branches of the orange-tree. These lice insert their beaks into the leaves and succulent twigs and live upon the sap. When full grown, they are a little more than one-twentieth of an inch long, black or brownish black, with plump, round bodies, long, yellowish antennæ, and pale-yellow legs. (See Fig. 398, where they are shown magnified.) The winged

Fig. 398.

specimens, one of which is seen in the figure, are also black; these fly from one tree to another and establish new colonies.

Remedies.—Syringe the trees with strong soap-suds or other alkaline washes, or with strong tobacco-water. A number of lady-birds and their larvæ, also the larvæ of Syrphus flies, feed on these lice. Many of them are destroyed by a minute Chalcid fly, which lives within their bodies. This friendly species, *Stenomesius aphidicola* Ashmead, is shown, much magnified, in Fig. 399, where *a* represents the female, and *b* the male. The short lines at the sides indicate their natural size. They are so minute that as many as three of the perfect winged flies have been known to issue from the body of a single aphis.

A tiny Ichneumon fly, the red-legged Trioxys, *Trioxys testaceipes* Cresson, also infests this species of aphis, while a third friendly parasite is a small Aphidius, a shining, black

FIG. 399.

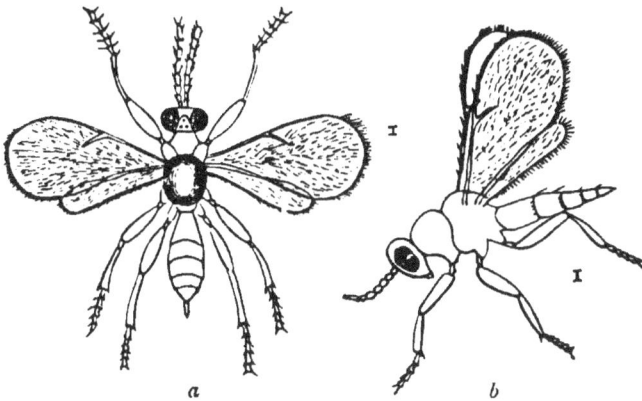

fly. Were it not for these predaceous and parasitic insects, the Aphides would soon multiply to such an extent as to ruin the plantations.

No. 249.—The Rust Mite.

Phytoptus oleivorus Ashmead.

The rust which often occurs on the fruit of the orange was until of late regarded as due to a fungoid growth, but recent investigators have shown that it is caused by a very small, four-legged mite, which punctures the oil-cells, and the exuding oil, when exposed to the influence of the atmosphere, soon undergoes a change, assuming a dark, rusty appearance, which seriously depreciates the value of the fruit for market. To the unaided eye the oranges appear dusty, but if examined with a magnifying-glass they will be seen covered with a multitude of mites of a whitish-flesh color.

A weak alkaline wash applied to the fruit would doubtless destroy these mites.

Another rust, known as "the black smut," often spreads

over both leaves and fruit, making them appear very unsightly. This is a minute fungous growth, known under the name of *Fumago salicina* Farlow, but it is believed by some to result from the punctures of insects, causing an exudation, on which the fungus thrives.

As a remedy, use an alkaline solution of soap as strong as the tree will bear without injury.

No. 250.—The Purple Scale.

Mytilaspis citricola Packard.

This is one of the most common and injurious species of scale-insect found in Florida. It is confined mainly to the

Fig. 400.

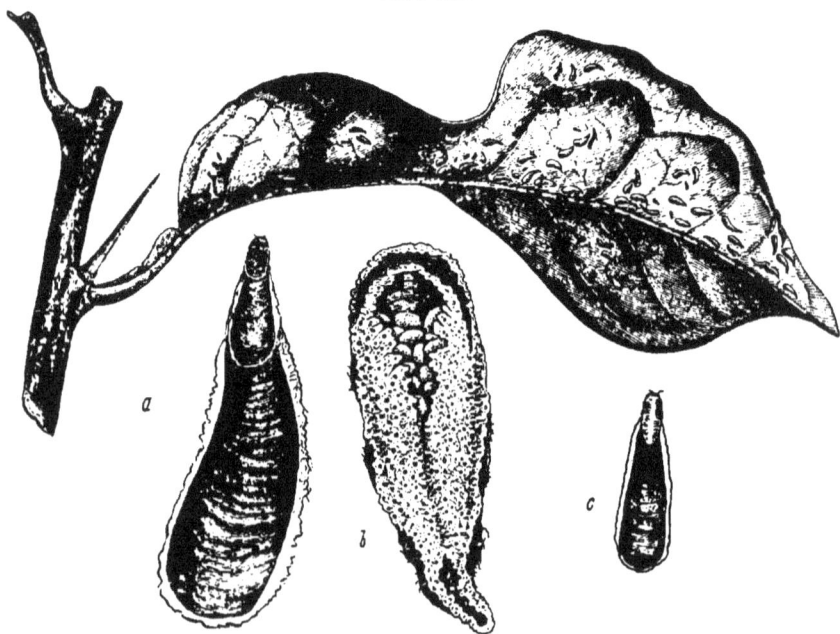

leaves and fruit of the orange, and sometimes disfigures the latter to such an extent as to make it unfit for market, yet it is often seen on fruit offered for sale. The scale of the female is shown empty at *a* in Fig. 400, and occupied by the insect at *b*, both highly magnified. It is long, narrow, more or less curved and widened posteriorly, varying in color from dark purple to reddish-brown, the enclosed insect being yellowish white. That of the male, shown at *c*, also magnified, resembles the female scale in form, but is nearly straight, and may be at once distinguished by its smaller size. In color it is much the same as the female scale, but is sometimes darker, occasionally dark brown or almost black. On the leaf in the figure these scales are shown of the natural size.

The eggs, which number from eighteen to twenty-five under each scale, are white, and are arranged irregularly, as shown at *b*. They hatch in Florida about the middle of March, producing lice of the form shown at *b* in figure 401, but so small as to be scarcely visible without a magnifying-glass. They are of a white color, yellowish at both ends, and have red eyes. For a very brief period after hatching they are active; then they fix themselves to one spot, where they remain stationary for the rest of their lives. Within a few days there is secreted over the body of the young louse a covering of fine cottony filaments, which, together with the skins shed from time to time as the insect increases in size, are eventually formed into scales, as shown in the figure. The male develops into a winged fly (see *a*, Fig. 401) which is red, with long, hairy antennæ and transparent wings; but the female remains within the scale and dies there.

FIG. 401.

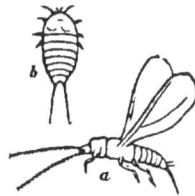

This scale-insect is said to have been imported from Bermuda on some lemons sent to Florida. Besides the lady-birds and other predaceous insects which attack all scale-insects, and which will be referred to in detail under "Remedies," this one has some special foes. A small mite, *Tyroglyphus Gloveri* Ashmead, is very useful in destroying it. The eggs of the mite are laid in December, in clusters of two or three hundred each, on the under side of orange leaves, close to the veins; they are of a reddish-yellow color, and about one five-hundredth of an inch long. Early in the year there hatch from them tiny blood-red mites having six legs, and four oval black spots on the hinder part of the abdomen. In three or four weeks these transform to eight-legged mites of a paler shade of red, which is the mature form.

A small, four-winged fly, one-fiftieth of an inch long, described as "the blue yellow-cloaked Chalcid," *Signiphora fla-vopulliatus* Ashmead, has been found in considerable numbers destroying the eggs of this scale. Fig. 402 shows this fly, highly magnified. Its body is bluish black, with a yellow crescent-shaped patch behind the head; the wings are transparent and fringed with fine hairs.

Fig. 402.

No. 251.—The Long Scale.

Mytilaspis Gloveri Packard.

The second most common scale-insect on the orange-trees in Florida is the species now under consideration. It is closely allied to No. 250, but differs from it in that the female scale is much narrower, and generally of a paler color, its usual tint being pale brownish yellow, varying occasionally to dark brown. A back view of the female scale is shown at *a* in Fig. 403, a front view at *c*, while the male scale is represented at *b*,—all magnified; on the leaf and twig are shown many scales of the natural size. The female insect, under the scale, is of a light-purple hue, with the terminal segment yellowish. The eggs are white when first laid, but become tinged with purple before hatching; they are arranged regularly in a double row, as shown at *c* in the figure. The newly-hatched lice are purplish, and resemble No. 250, as shown at *b*, Fig. 401. They are active for a brief period, and then settle permanently in

FIG. 403.

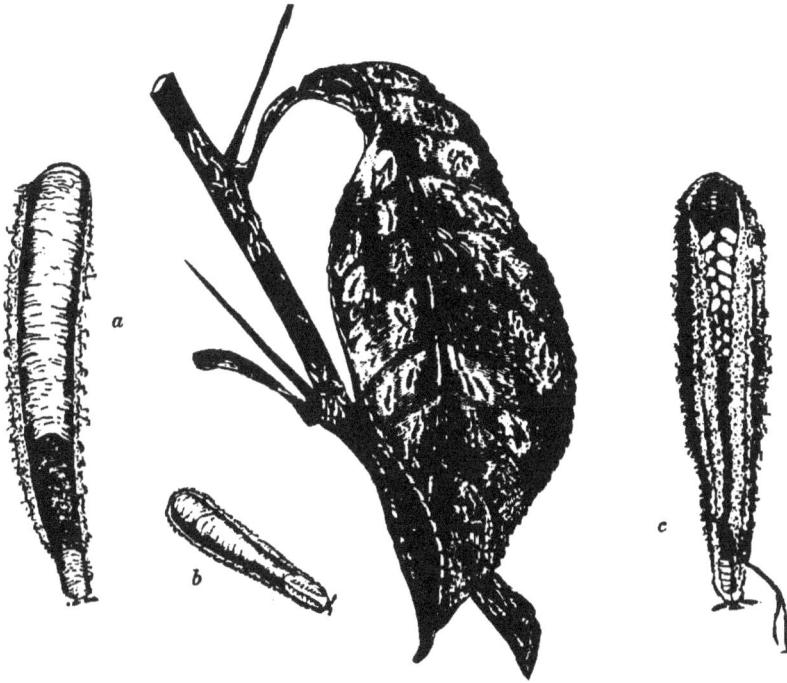

one spot, where they remain stationary. The male insect is a very minute fly, which is shown, highly magnified, in Fig. 404. It has long antennæ and two transparent wings.

This species is found on trees of the Citrus family throughout Florida, also in Louisiana, infesting the twigs and branches, and finally the leaves, but rarely the trunk. There are three broods in a season. It is said to have been imported from China, and has since been disseminated by the distribution of infested nursery stock and by the fruit itself.

This insect also has some special parasites; one, a tiny four-winged fly, *Aphelinus aspidioticola* Ashmead, is about one-fiftieth of an inch long, of a light-brownish color, with fringed wings. (See Fig. 405.) It lays an egg under each scale, the larva from which is a white, fleshy, footless grub,

FIG. 404.

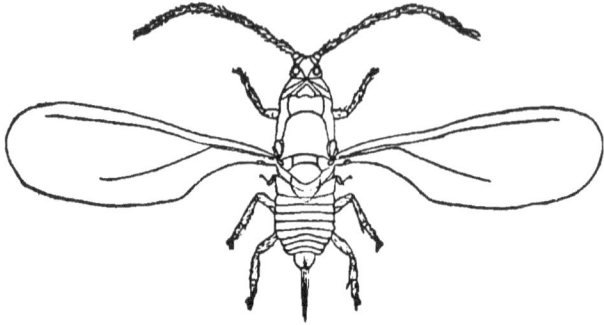

that feeds upon the eggs. By the time it has consumed them all it has reached full growth, when it changes to a pupa, and, after remaining in this condition a few days, the fly escapes by eating a passage through the top of the scale. Where this parasite does not occur, it may be introduced with advantage by taking into the locality branches infested with scales which are known to have been parasitized. This useful insect destroys immense numbers of the scales, and is doubtless one of the chief natural agencies provided to check their undue increase.

FIG. 405.

A species of mite, *Oribates aspidioti* Ashmead, has been found feeding on the eggs of this scale-insect. It is about one-fiftieth of an inch long, of an elongated, flattened form and a dark reddish-brown color.

No. 252.—The Red Scale of California.

Aspidiotus aurantii Maskell.

The female scale of this species is quite translucent, its
apparent grayish color depending on that of the insect

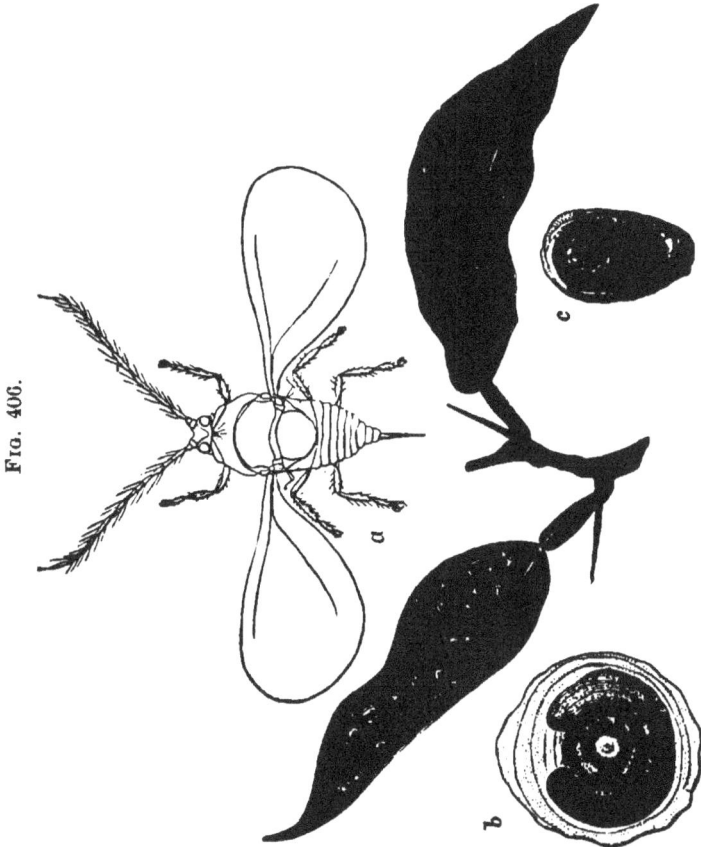

Fig. 406.

beneath, which varies from a light greenish yellow to a
bright reddish brown, and when the female is fully grown
the form of its dark body shows distinctly through the
transparent covering, as represented at *b* in Fig. 406. The
scale of the male, shown at *c* in the figure, resembles that of

mature rarer Irlysten of Trost.
"Chilocorus Similis"

the female, but is only one-fourth the size, the posterior side being prolonged into a flap, which is quite thin. The scales are represented of the natural size on the leaf and twig. The perfect male insect, which is winged, as shown, highly magnified, at *a* in Fig. 406, is light yellow, with a brown band on the thorax, and purplish-black eyes. The eggs are of an ovoid form and bright-yellow color, from twenty to forty being found under each scale.

This species appears to confine itself to the trees belonging to the Citrus family, and infests the trunk, limbs, leaves, and fruit, sometimes covering the latter to such an extent as to render it unfit for market. Where these insects are very numerous, the leaves turn yellow, and sometimes drop from the trees. In Southern California there are five or six broods during the year; hence it is spreading with great rapidity, and is perhaps more to be dreaded than any other scale-insect in this country. Many groves in Los Angeles and in other sections of Southern California have been seriously injured by it. The orange-groves in Australia have suffered from the same pest.

No. 253.—The Circular Scale.

Aspidiotus ficus Riley.

This is known as the red scale of Florida. In Fig. 407 the scales are shown of the natural size on the leaves of an orange-tree; *a*, the scale of the female; *b*, that of the male; *c*, the young larva; *e* and *f*, different stages in the formation of the scale; all these are highly magnified. Thus far it has been found only in the orange-groves of Florida. It multiplies with great rapidity, and infests indiscriminately the limbs, leaves, and fruit.

The scale of the female (*a*) is circular, and varies from a light to a dark reddish-brown color, with a gray margin; that of the male (*b*) is about one-fourth the size of the female scale, and of a dark reddish brown, with a white centre, and is prolonged into a thin flap, of a grayish color.

The eggs are pale yellow, and the newly-hatched larvæ, shown at *c* in the figure, are broadly oval in outline, and are each provided with six legs, a pair of antennæ, and a beak

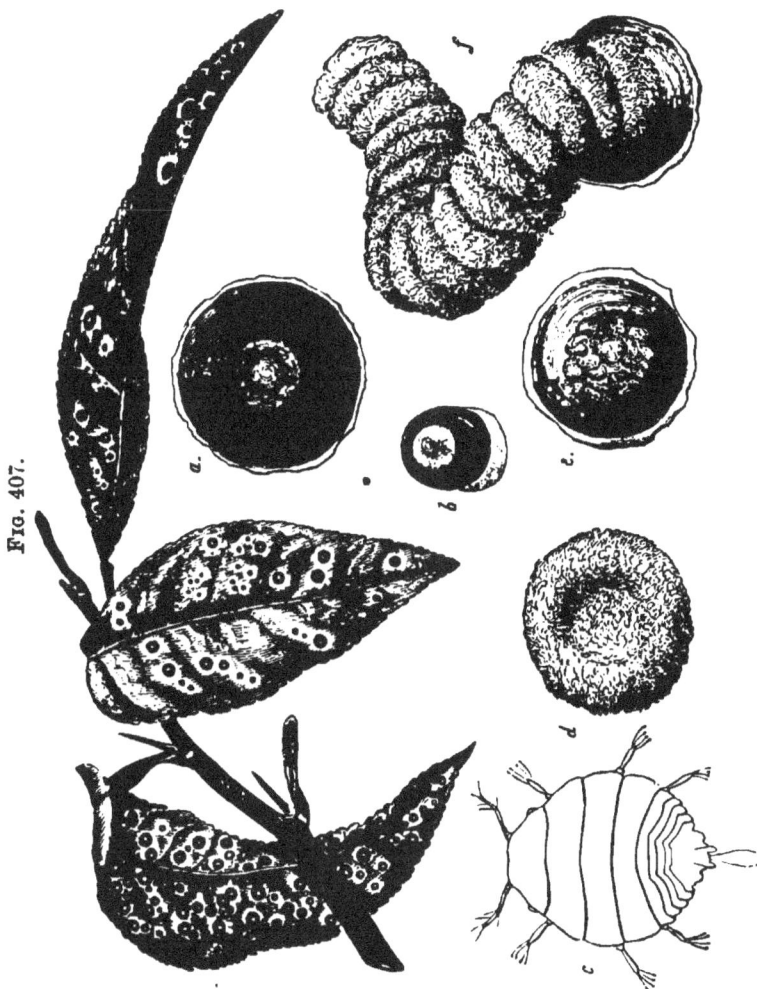

FIG. 407.

for suction. They appear as small specks, scarcely visible to the unaided eye; at first they are quite active, but, having selected a location, soon fix themselves permanently to one spot. In a short time they secrete over their bodies fine threads of wax, which are cottony in appearance. Soon a small, white, convex scale takes the place of this cottony coating, which is depressed in the centre. (See *d*, Fig. 407.) The scales gradually increase in size, and as they approach maturity there is secreted on the female scale a mass of cottony threads, which increases in quantity until it some-

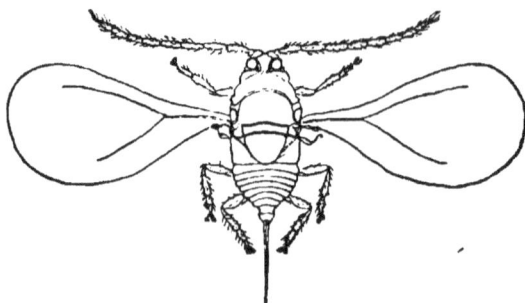

times extends in a curved form, as shown at *f*, to a length five times the diameter of the scale. In the figure all the illustrations are highly magnified, except the leaves with the scales on them, which are of the natural size.

The male is furnished with a single pair of large, transparent wings, which enable it to fly readily. It is shown, highly magnified, in Fig. 408.

No. 254.—The White Scale.

Aspidiotus nerii Bouché.

This scale is found on the orange and lemon trees, particularly in Southern California and in Florida, where it also infests a number of other trees and plants, but especially the acacia-tree. In Fig. 409 a twig of acacia is figured infested with this scale. The female scale is flat, whitish or light gray in color, and when mature is only about one-twelfth of an inch in diameter. The eggs are of a light-yellow color. The scale of the female is shown at *c* in the figure; the male

scale at *b*, both magnified ; the latter is slightly elongated
in form, of a white color, with a tinge of yellow, and is about
one twenty-fifth of an inch in diameter.

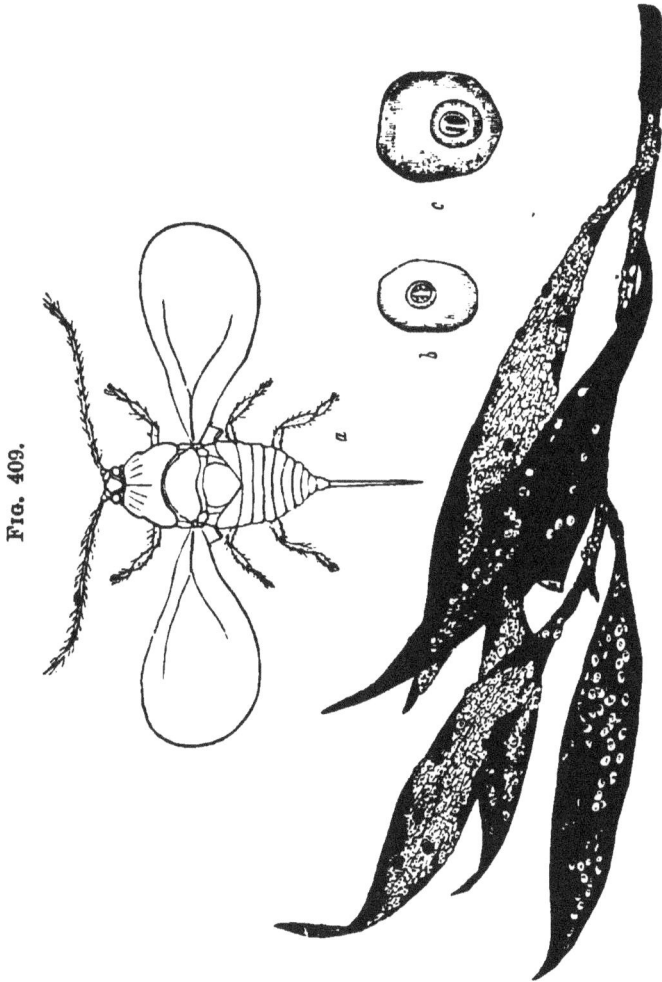

The winged male, which is a very minute creature, is
shown, highly magnified, at *a* in the figure; it is yellow,
mottled with reddish brown ; wings transparent.

No. 255.—The Ribbed Scale.

Icerya purchasi Maskell.

The adult female of this species of Coccus is covered by an egg-sac, which is of a pale-yellowish color, longitudinally ribbed, a little longer than the body of the insect, and filled with a loose, white, cottony matter containing the eggs. A cluster of these sacs is shown in Fig. 410, of the natural size;

Fig. 410.

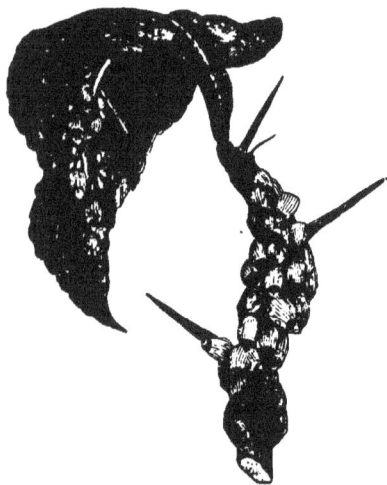

the enclosed insect is of a dark orange-red color, with black antennæ and legs, its back being covered more or less with a white or yellow-ish-white powder.

The eggs are said to number from two hundred to five hundred in each cluster, and are of a pale-red color. The newly-hatched larva is reddish or brownish, with long and slender legs. As it grows it gradually changes, becoming darker in color and irregular in outline, and it soon begins to excrete tufts of waxy matter along the back and sides, following which long, semi-transparent filaments appear.

These insects first attack the leaves, usually along the midrib, and afterwards migrate to the twigs and branches, and sometimes attach themselves to the trunk. They spread with amazing rapidity on orange and lime trees, the trunks and limbs of which are sometimes so completely covered with them as to appear white; the leaves turn yellow and sickly, and if no remedial measures are adopted the trees sometimes die. The insect has been found very destructive at Santa Barbara, where it has probably been introduced with plants from Australia.

No. 256.—The Chaff Scale.

Parlatoria Pergandii Comstock.

In this species the scale of the female varies in form, being sometimes nearly circular, but more usually somewhat elongated, of a dull-gray color, and thin in its structure. It resembles the bark so closely in tint that it often escapes detection. In length it is about one-sixteenth of an inch; the enclosed insect is nearly as broad as long. These insects vary greatly in color, some being almost white, with the extremity of the body slightly yellow; others are entirely yellow, while some are purplish, with the end of the body yellow. The eyes are black.

Fig. 411.

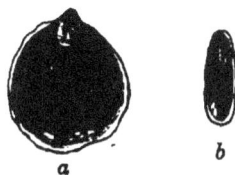

a *b*

Scales of both sexes are shown, magnified, in Fig. 411, *a*, *b*. The eggs and young larvæ are purplish. The scale of the male (*b*) is about one twenty-fifth of an inch long, and narrow; its color is gray, darker and greenish about the middle.

The mature winged insect is shown in Fig. 412, much magnified; it is purplish in color, with the disk of the thorax pale and irregularly marked with purplish spots. The eyes are large and very dark. There are several broods of these insects during a season, and the scales may

Fig. 412.

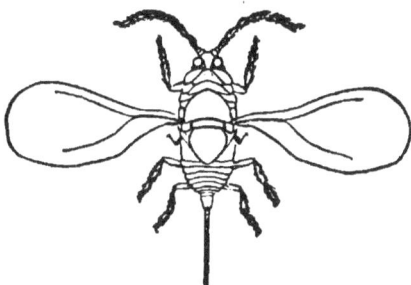

be found at any time on the bark of the trunk and branches of the orange-trees, and to a less extent on the leaves and fruit. They have been called chaff scales, from their resemblance to fine chaff or bran.

No. 257.—The Barnacle Scale.

Ceroplastes cirripediformis Comstock.

The color of this scale varies from grayish to light brown,

Fig. 413.

divided by lines into regular segments, as shown at *a* in Fig. 413, where one of these scales is represented magnified. The enclosed insect is subglobular in form, and of a dark reddish-brown color.

The eggs are light reddish brown, and rather long and slender; the larva is dark brown, and very slender in form. It is at first active for a brief period, then settles in one spot, where it becomes stationary, and soon secretes over its body tufts of cottony filaments, which are finally condensed to a waxy consistence, forming part of the scale with which the insect is covered.

This scale is found in several localities in Florida on both orange and quince trees; it is also found on a native plant, a species of Eupatorium.

No. 258.—The Florida Ceroplastes.

Ceroplastes Floridensis Comstock.

This scale is at first white; afterwards it becomes pinkish, growing redder or brownish in the middle, dull white towards the edges, some specimens being irregularly mottled with brownish and yellowish white, the top ornamented with

lines and dots, as shown at *b* in Fig. 414. The eggs, which often number a hundred under a single scale, vary in color from yellow to light reddish brown, and are nearly oval in form. The young louse is of a similar color, very active, and when first hatched appears as shown in Fig. 415, where it is much enlarged. It crawls about briskly for half an hour or more, then settles

FIG. 414.

FIG. 415.

on some spot, inserts its proboscis, and remains permanently fixed. Within a few days the limbs are drawn under the body, and white, cottony tufts are secreted from the surface; these gradually condense, forming waxy plates, which cover and protect the insect beneath. The scales are shown of their natural size, on a branch of ilex, in Fig. 414; a young female scale is shown at *a*, and a mature one at *b*, both enlarged.

This scale is common on the orange, lemon, and other trees

of the Citrus family in Florida; also on the fig, pomegranate, guava, quince, Japan plum, red bay, oleander, and sweet bay, and is very abundant on the gall-berry, *Ilex glabra*. It is referred to in W. H. Ashmead's "Treatise on Orange Insects" under the name of the white scale, *Ceroplastes rusci* Linn. There are three broods during the year: the first appear in April and May, the second from the middle to the end of July, and the third during the first two weeks in September. They increase with marvellous rapidity, but are preyed on by a species of Chalcid fly and by other insect enemies.

No. 259.—The Broad Scale.

Lecanium hesperidum Linn.

Fig. 416.

Of all the bark-lice here treated of, few are so common, and none so widely distributed, as this species. It is found in abundance from Washington southward to Florida, also in Utah and California, on the twigs of orange and other trees, shrubs, and plants; but, having so many different food-plants, it is not so destructive to the orange as are some others which confine their attacks to trees of the Citrus family. The scale is brown, sometimes quite dark, and is represented of its natural size on the stem of the twig in Fig. 416. It is one of the largest scales found on the orange; it is of an elongated, oval form, and highly convex. The enclosed insect is yellow, inclining to brown, of an elongated, oval form, nearly flat, smooth, and shining.

The young larva (see Fig. 417) is of a long, oval form, of a yellowish color, with two long thread-like filaments extending from the hind segment.

FIG. 417.

This bark-louse is much infested by parasites, no less than three distinct species having been bred from the scales.

The first of these, *Coccophagus cognatus* Howard (see Fig. 418), is a very small, four-winged fly, the female of which, when its wings are spread, measures about one-twelfth of an inch, the male about one-sixteenth. The

FIG. 418.

FIG. 419.

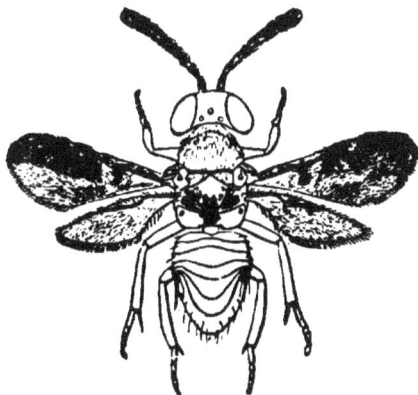

body is of a dark-brown color, with yellow markings; the wings are transparent.

In Fig. 419 is shown another of the parasites of this scale-insect, known as *Comys bicolor* Howard, a small fly, which measures, when its wings are expanded, nearly one-eighth of an inch across. The fore wings are dusky brown on their outer two thirds, the inner portion nearly transparent, with a brownish streak; the hind wings are nearly transparent. The body is black, the thorax brown, with black hairs. This insect has been found very abundant in Washington, destroying large quantities of the broad scale-insects which occur in multitudes on the English ivy grown there.

Both sexes of a third parasite, *Encyrtus flavus* Howard, are shown in Fig. 420, *a* representing the male, *b* the female.

Fig. 420.

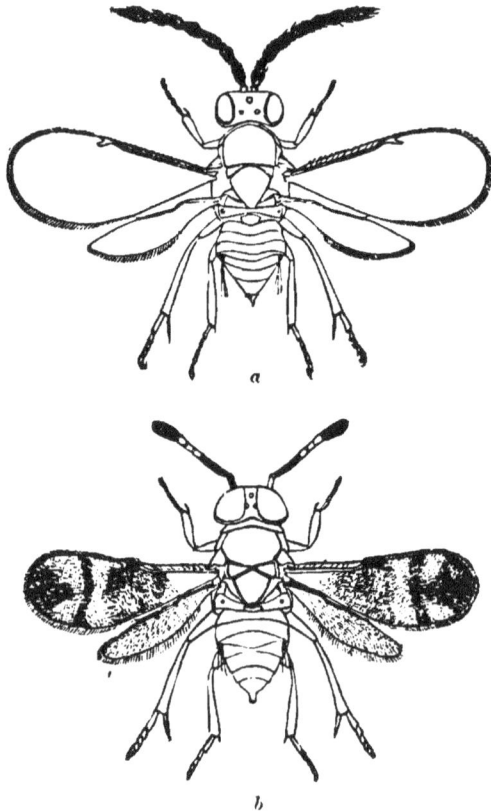

The wings of the former measure, when spread, about one-eighth of an inch; those of the latter, one-tenth of an inch. The basal third of the fore wings of the female is transparent, the middle third dusky brown, crossed by a clear transverse band; the outer third is also dusky brown, with two large, wedge-shaped, transparent spots entering it, one from each side. The hind wings are nearly transparent; the body is ochre-yellow, with brown markings. The male is of a shining metallic-green color, with yellow markings; the wings are transparent. This parasite has been bred from orange-trees in Southern California. All these parasites are shown highly magnified.

No. 260.—The Black Scale of California.

Lecanium oleæ Bernard.

In France, where this scale is also found, it chiefly affects the olive-tree, but in California it has been found on a great variety of trees, and has become a serious enemy to

Fig. 421.

orange-culture, being perhaps more generally distributed on the orange-trees in that State than any other species of scale-insect. Besides the orange, lemon, and other members of the Citrus family, it is found on the olive, pear, apricot, plum, pomegranate, apple, and a number of other trees, shrubs, and plants. The scales are usually found on the smaller twigs. In Fig. 421 they are shown, of the natural size, on an olive-twig; and at *a* in the same figure a scale is shown mag-nified. The scales are blackish brown, marked with ridges and indentations, as indicated in the fig-ure. The eggs are of a long, oval form and yellow color. The male, though diligently sought for, has not yet been discovered.

FIG. 422.

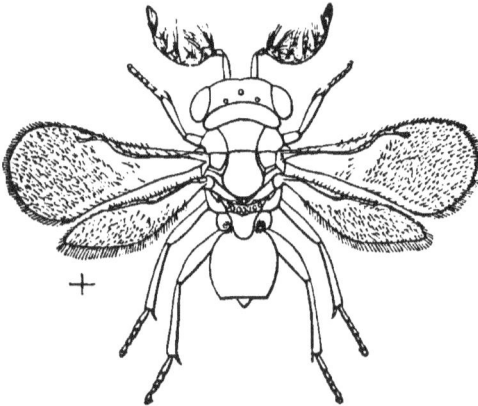

In Fig. 422 is shown the male, and in Fig. 423 the female (both en-larged), of a very interesting little fly, *Tomocera Califor-nica* Howard, which is a parasite on this black scale. The wings, which are transparent in both sexes, measure, when spread, a little more than one-eighth of an inch across. Its general color is deep blue-black, with a metallic lustre and brown markings. The male may be distinguished from the female by its shorter body and peculiar antennæ. This para-site is so abundant in some sections that as large a proportion

FIG. 423.

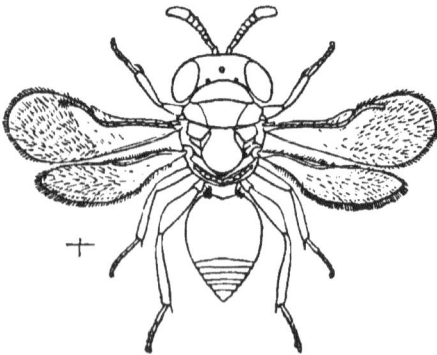

as seventy-five per cent. of the scales have been known to be destroyed by it. The female fly pierces the scale and deposits in it a single egg. When hatched, the larva feeds upon the eggs and young of the bark-louse, and later upon the mother also. When full grown, it is nearly one-sixth of an inch long, broad, becoming narrower towards the head, of a transparent white color tinged with blackish from the alimentary canal showing through. The larva changes to a pupa within the scale, which at first is white, but soon becomes darker in color; the fly, on escaping, makes its exit through a round hole which it cuts in the back of the scale.

No. 261.—The Hemispherical Scale.

Lecanium hemisphæricum Targioni.

Fig. 424 represents this scale, of its natural size, on orange leaves, and a magnified one at *a*. It varies in color from light to dark brown, and is occasionally tinged with reddish when mature. In shape it is hemispherical, with the edges flattened, its form varying somewhat in different situations; upon a rounded twig it becomes less hemispherical, more elongated, and its flattened edges are bent downwards, clasping the twig.

The eggs are yellowish white, smooth, and shining. The newly-

FIG. 424.

hatched larvæ are very active, and even the adult insect can crawl from one point to another with apparent ease, carrying the scale with it.

This scale has been found on orange-trees near Santa Barbara, and doubtless exists in other localities also. In greenhouses it attacks not only the orange but many other plants.

No. 262.—The Common Mealy-bug.

Dactylopius adonidum Linn.

The insects known under the name of mealy-bugs form no scale, and are not always stationary, having the power of moving from one place to another; but, since they require the same treatment as scale-insects, it will be convenient to treat of them here. This species of mealy-bug is common in green-houses throughout the civilized world. The female is represented magnified in Fig. 425, with most of the mealy matter removed. When full grown, it is about one-eighth of an inch long, white, with a tinge of yellow, a brown band upon the middle of the back, and its whole body powdered with white, floury-looking material. The sides and extremities of the body are armed with spines. The larva, which varies in size according to its age, is of the same form, but flatter.

Fig. 425.

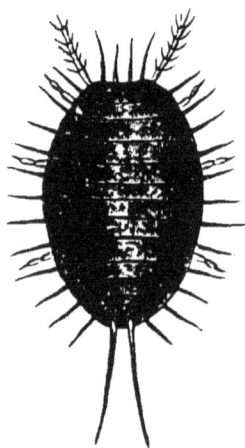

The male is a small winged insect, much resembling that of No. 263.

In Florida it attacks the orange, guava, grape-vine, and pineapple, and prevails to such an extent that it is said few orange-trees have escaped its ravages except those in the interior and southern parts of the State.

No. 263.—The Destructive Mealy-bug.

Dactylopius destructor Comstock.

The name *destructor* has been proposed for this species of mealy-bug on account of the injury done by it to orange-trees in Florida, where it is one of the most serious insect pests with which the orange-grower has to contend. The adult female, which is shown magnified in Fig. 426, is about one-sixth of an inch long, and half that in width, and has seventeen lateral appendages on each side, which are nearly uniform in length. There is a slight powdery secretion distributed over the body. The female begins laying her eggs in a cottony mass at

Fig. 426

the extremity of the abdomen before she attains full growth, and the egg-mass increases with her growth, gradually forcing the hinder portion of the body upwards, until finally she appears as if almost standing on her head.

The eggs are rather long, and of a bright straw-color, and, soon after hatching, the young larvæ, which are rather brighter in color than the egg, spread in all directions, settling preferably along the midrib, on the under side of the leaves, or in the forks of the young twigs, where they form large colonies, closely packed to-

Fig. 427.

gether. The young are only slightly covered with white powder.

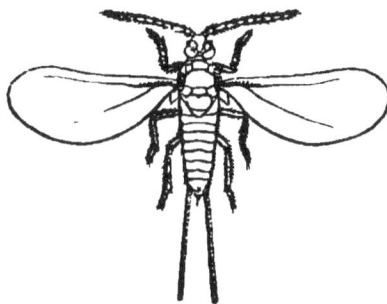

The male, which is represented highly magnified in Fig.

427, is furnished with two transparent wings, which, when spread, measure rather less than one-eighth of an inch across.

Its body is olive-brown; the eyes are dark red.

Fig. 428.

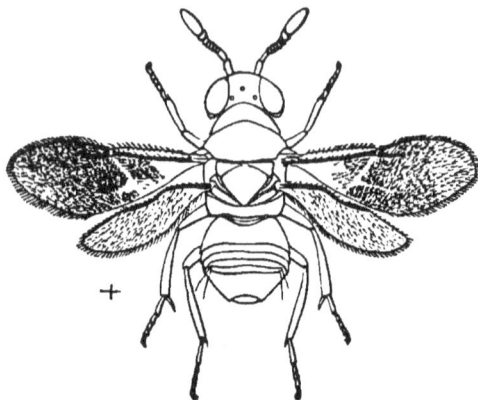

The four-winged fly shown, much magnified, in Fig. 428, the natural size of which is indicated by the short lines on the left of the figure, is a parasite on this mealy-bug, known as *Encyrtus inquisitor* Howard. Its body is smooth, of a shining black, and the transparent wings are partly obscured by dusky markings, as shown in the figure.

No. 264.—The Mealy-bug with Long Threads.

Dactylopius longifilis Comstock.

In this species the adult female is nearly one-fifth of an inch long, of a light dull-yellow color, its body being covered with a whitish powder. In Fig. 429 it is represented magnified. The lateral appendages, which are seventeen in number, are long, the posterior ones on each side being very long, equalling, and sometimes exceeding, the entire length of the body. In the larval state the male and the female are very much alike, but as they approach maturity striking differences appear. The female surrounds herself with cottony material, amid which the young cluster for some time after birth. The male larva forms for itself a little cottony sac or cocoon, in which it changes to a pupa, from which the winged insect is produced. This is shown, much magnified, in Fig. 430. The wings, which are transparent, measure, when spread, about one-tenth of an inch across. The body is brown; the eyes are dull red.

Fig. 429.   Fig. 430.

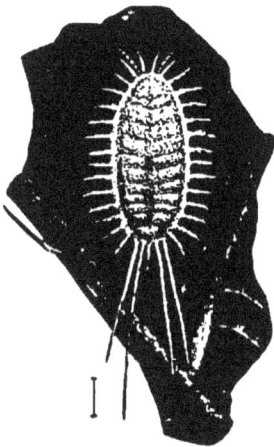   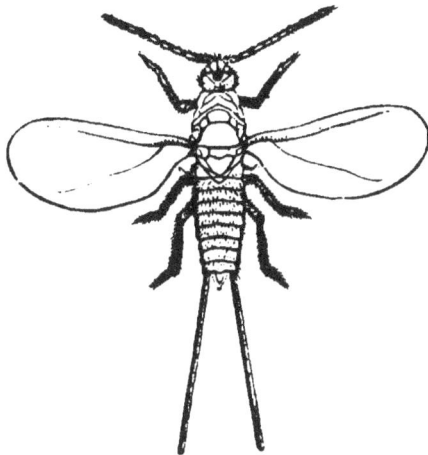

REMEDIES.

In treating of the remedies for scale-insects and mealy-bugs, those provided by nature will first claim our attention. Under the several species discussed, reference has been made to the parasitic flies which destroy them, as these are often limited in their attacks to one species. The predaceous insects, which feed on them indiscriminately, will now claim attention; these consist mainly of various species of lady-birds. These useful insects vary in size, and are usually red, yellow, or black, with spots of one or the other of these colors. Some of them are found from the Atlantic to the Pacific, such as the nine-spotted lady-bird, Fig. 123; the plain lady-bird, Fig. 125; the convergent lady-bird, Fig. 128; the spotted lady-bird, Fig. 129; and the twice-stabbed lady-bird, Fig. 33. Those which follow are restricted to the Pacific coast, or are more abundant there. Lady-birds, both in their larval and in their perfect state, devour scale-insects, mealy-bugs, and aphides.

The Ashy-gray Lady-bird.

Cycloneda abdominalis (Say).

This is a small-sized lady-bird, which is often found in abundance on infested orange-trees. Its larva also is very common, and, when full grown, measures about four-tenths of an inch long. It is black, variegated with orange, yellow, and greenish white, and is shown, magnified, at Fig. 431, *a.*

When about to transform to a pupa, the larva attaches the end of its abdomen to a leaf, when shortly the skin, splitting at the back of the head, gradually shrivels up towards the posterior end, revealing the pupa, as shown in the figure at *b.* This is of a whitish color, tinged in some parts with yellowish, and ornamented with black spots.

Fig. 431.

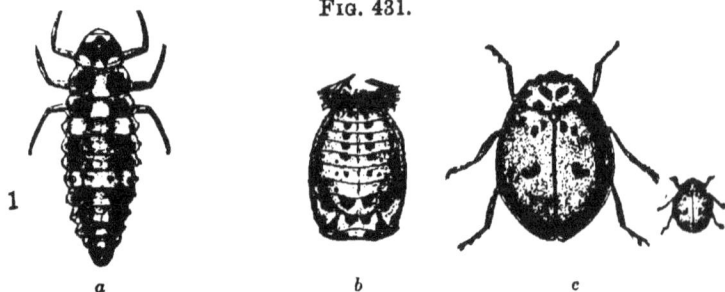

The beetle is ashy gray, with seven black spots on the thorax, and eight upon each wing-cover, arranged as shown at *c* in Fig. 431, where the insect is represented magnified, the smaller figure at the side indicating the natural size.

The Blood-red Lady-bird.

Cycloneda sanguinea (Linn.).

The blood-red lady-bird is not so common as the species last described, but is nevertheless very useful. The larva is without spines, flattened in form, and ornamented with transverse yellow bands and black spots; it is most common in the spring, when it is exceedingly voracious and active.

The pupa is shown magnified at *a*, in Fig. 432. It is about a quarter of an inch long, of a broad, oval form, and of a dull-yellow color, with orange and black markings.

The beetle, which is represented magnified at *b*, and of the natural size at *c*, in the figure, is almost hemispherical in form, and red, varying in the depth of its hue from a pale-red to a blood-red color. The thorax is

FIG. 432.

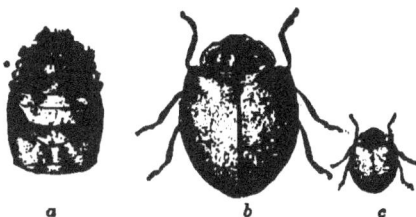

a   b   c

black, with its margin and two spots of an orange color, the head black, with two pale spots. This species has already been referred to under the name of the plain lady-bird (Fig. 125), under which designation it has long been known in the East.

The Cactus Lady-bird.

Chilochorus cacti (Linn.).

This beetle is also known to destroy scale-insects. The larva is shown, magnified, at *a* in Fig. 433. It is black,

FIG. 433.

a   b   c   d

crossed by a light-yellowish band about the middle, and is armed with many long, branching spines. The pupa, also magnified, at *b* in the figure, is formed within the larval skin, which splits open along the back sufficiently to show the enclosed pupa, which is black, with a few sparsely-scattered tufts of fine hair.

The beetle, which is seen magnified at *c*, and of the natural size at *d*, is of a shining black color, with an irregular reddish spot on each wing-case, and much resembles the twice-stabbed lady-bird of the East.   (Fig. 33.)

The Ambiguous Hippodamia.

Hippodamia ambigua Lec.

In many districts in California this is a very abundant insect.   The larva is shown in Fig. 434 at *a*, and, when full

FIG. 434.

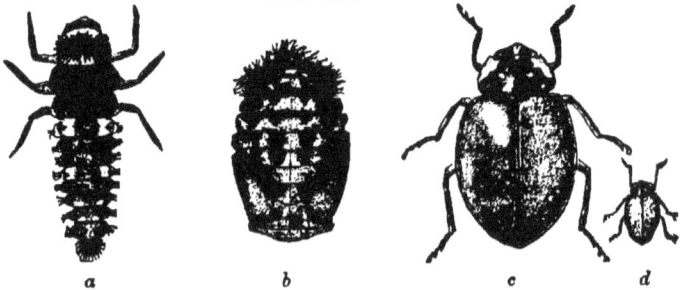

a   b   c   d

grown, is about half an inch long, of a bluish-black color above, marked with orange, black, and yellowish white. The pupa, *b*, is nearly one-third of an inch long, of a dull orange-yellow, with black and yellow markings. The beetle, *c, d*, resembles the blood-red lady-bird, but is narrower in proportion to its length, and less convex in form.   The head is black, with a whitish patch in front, and the thorax black, with a dull-white patch on each side towards the front.   In the figure, *a, b*, and *c* are magnified, and *d* shows the natural size.

The Eyed Cycloneda.

Cycloneda oculata (Fab.).

This species, which is represented magnified at *a*, Fig. 435,

FIG. 435.

a   *b*

and of its natural size at *b*, has black wing-covers, with a large reddish spot on each.

The Five-Spotted Lady-bird.

Coccinella 5-notata var. *Californica* Mann.

Fig. 436 shows the Californian variety of the five-spotted lady-bird, which is a form with no spots. The thorax is

FIG. 436.

a   *b*

black, with a pale spot on each side, and the wing-covers pale orange.

27

In addition to the species already named, the following are worthy of mention :

Exochomus contristatus Muls. This is a small lady-bird, about one-seventh of an inch long, of a red color, with a black thorax and two black spots on the wing-covers, placed near the hinder end. The larva is about one-sixth of an inch long, yellowish, with black spots and spines. Both the larva and beetle are useful in destroying scale-insects, and are quite common among the orange groves.

Scymnus cervicalis Muls. A hemispherical beetle, about one-tenth of an inch long, of a reddish-brown color, with dark-blue wing-covers. Its larva is pale whitish, with a few scattered hairs, the head small, round, and black.

Scymnus bioculatus Muls. The larvæ of this beetle have been found feeding on the eggs of the mealy-bug; they are covered with a white secretion, something like the mealy-bug itself, and hence are not easily discovered.

Hyperaspidius coccidivora Ashmead. This beetle, which resembles a minute Scymnus, also destroys many of the scale-insects, and is especially destructive to the chaff scale. It is about one twenty-fifth of an inch long, oval, of a dark color, having a polished surface and a reddish patch on each wing-cover.

Fig. 437.

The orange Chrysopa, *Chrysopa citri* Ashmead. This is a lace-wing fly, of a bright yellowish-green color, with antennæ longer than the transparent, netted wings, and having bright, golden eyes. (See Fig. 437.) Its eggs are laid on long, thread-like stalks, and the larva, which devours both scale-lice and plant-lice greedily, covers itself with minute pieces of dried leaves or other light substances. It is pinkish, mottled with brown spots.

Artificial Remedies.—From the suctorial habits of the bark-lice, the remedies available are limited to such as destroy life by contact, or produce death when inhaled through the breathing-pores; for since these insects draw their food from beneath the surface of the tissues, the application of any poison which requires to be eaten with the food to produce its effects is not likely to be of much service.

Scale-insects on the bark of the trunk or limbs of trees may be removed mechanically by using a stiff brush, either with or without the use of an insecticide. Those on the smaller twigs and leaves can only be reached by spraying some suitable liquid on the trees. Alkaline washes seem to have successfully stood the test of practical experiment, and are used with good results by many of the leading fruit-growers on the Pacific coast and in Florida.

A solution of concentrated lye or commercial potash, or its equivalent in lye made directly from wood-ashes, appears to be equally effective.

One bushel of good wood-ashes will produce about four pounds of potash; hence, in making alkaline washes for trees, this estimate may be acted on where concentrated lye cannot be conveniently procured. To obtain the potash in solution, place a bushel of ashes in a keg or barrel having a tap or spigot near the bottom. Press them firmly and evenly down, and lay a small piece of board on the ashes, so that the water, when poured on them, shall not disturb their surface. Pour hot water on the board, so that it may spread and soak evenly through the ashes, using a sufficient quantity to saturate them thoroughly. Allow it to stand twenty-four hours, then draw off the lye at the tap, adding more water to displace that held by the ashes, until eight gallons are obtained. As the first portion of the liquid which comes off will be much stronger than the last, agitate the solution so that it may be thoroughly mixed. Each gallon may then be estimated to contain half a pound of commercial potash.

For cleansing orange or other Citrus trees from scale-insects, take one pound of concentrated lye to three gallons of water, or one and a quarter pounds of commercial potash, or its equivalent, ten quarts of the home-made lye, and make the solution up to three gallons with water. Before the trees bloom, thin out the branches by pruning, so that air and light may have free access to the foliage and fruit, carefully burning all the prunings; then wash or spray the entire tree, trunk, limbs, and foliage, and, if practicable, use the wash heated to a temperature of about 130° F., which would be nearly as hot as the hand could bear.

In two or three weeks, or about the time when the young larvæ appear, the washing or spraying should be repeated, using the same mixture, but adding to each gallon half a pound of flour of sulphur; or use a solution of whale-oil soap, containing from one-quarter to three-quarters of a pound to the gallon, with half a pound of sulphur. If the insects are not entirely subdued, after an interval of three or four weeks a third application may be made. If the trees require treatment while in bloom, it is safer to use the soap solution, as the stronger alkaline washes sometimes injure the tender growth. For scales on apple, pear, plum, cherry, peach, apricot, and nectarine trees, the solutions may be used one-third stronger, but may be made twice the ordinary strength when applied with a brush to the trunk and limbs only.

During the earlier period of their growth, scale-insects are readily destroyed by insecticides of moderate strength, especially while in the active larval stage, but when the tough scales are well formed they are much more difficult to exterminate. While reproduction to some extent appears to be going on from March to December with but little cessation, there is no doubt that the months of March, June, and September mark the appearance of a very large proportion of the successive broods; hence, during these months, remedies can be applied with the greatest advantage. Those pests which are unprotected by scales, such as the mealy-bugs, can be destroyed at any time with comparative ease by the use of the alkaline or soap solutions.

Strong tobacco-water, heated to about 130° F., has also been used with some success, more particularly on the young broods.

Diluted emulsions of kerosene oil are also valuable agents in destroying the different species of bark-lice, as well as many other injurious insects. Emulsions prepared in the following manner have been found very efficient in several series of experiments conducted under the direction of the Department of Agriculture at Washington :

No. 1.—Kerosene oil, ... 2 gallons.
Common soap, ... ½ pound.
Water, ... 1 gallon.

Dissolve the soap in the water and heat the solution, adding it, boiling hot, to the kerosene. Churn the mixture with a force-pump and spray-nozzle for five or ten minutes, when the emulsion, if perfect, forms a cream which thickens on cooling and should adhere without oiliness to the surface of glass. Dilute this emulsion with from 10 to 12 times its bulk of cold water, and spray it on the foliage.

No. 2.—Kerosene oil, ... 2 gallons.
Sour milk, ... 1 gallon.

Warm the ingredients to a blood-heat and emulsify in the same manner as is directed for No. 1, and subsequently dilute with from 10 to 12 parts of water before using.

No. 3.—Take the white of two eggs, three tablespoonfuls of sugar, a pint and a half of water, and two pints and a half of kerosene oil. Emulsify with a force-pump and spray-nozzle, when a cream-like compound will be produced, which should be diluted with from 10 to 12 times its bulk of water.

It is said that these diluted kerosene emulsions, when properly prepared, so that the oil does not separate, are more effective than the alkaline washes, and that they do not injure the trees.

For the application of these fluids several forms of portable pumps have been devised, in the selection of which the fruit-grower should be guided by his own requirements. Where the orchard is large, it will pay to purchase an efficient instrument for this purpose. It is stated that, with a suitable pump and nozzle for spraying, from one to two hundred trees can be thoroughly treated in a day.   •

Since by far the greater portion of the injury caused by insects to orange-trees is effected by the scale-insects, it is important that prompt measures be adopted to destroy them, and that every precaution be taken to prevent their introduction into districts hitherto exempt from them. Many localities have been colonized by these pests through the return of empty fruit-boxes from infested districts. These may be disinfected by dipping them for at least two minutes in boiling water containing not less than one pound of potash or half a pound of concentrated lye to each twenty-five gallons. These insects are also frequently disseminated by the transportation of nursery stock from one part of the country to another.

Sickly trees are more predisposed to attack than healthy ones; hence the use of fertilizers to induce a vigorous growth has been suggested as a remedial measure. In planting new groves, avoid the vicinity of diseased trees if possible, as the young lice are liable to be carried some distance by winds, or on the feet of birds visiting the trees.

No. 265.—The Greedy Scale-insect.

Aspidiotus rapax Comstock.

The scale of the female in this species is about one-sixteenth of an inch long, very convex, of a gray or drab color, and somewhat transparent. The enclosed insect is bright yellow, with translucent blotches. It is shown in the natural position on a limb, and also detached, in Fig. 438.

The eggs, which are found under the mature female scales, are yellow, so also are the newly-hatched larvæ; the latter

FIG. 438.   FIG. 439.

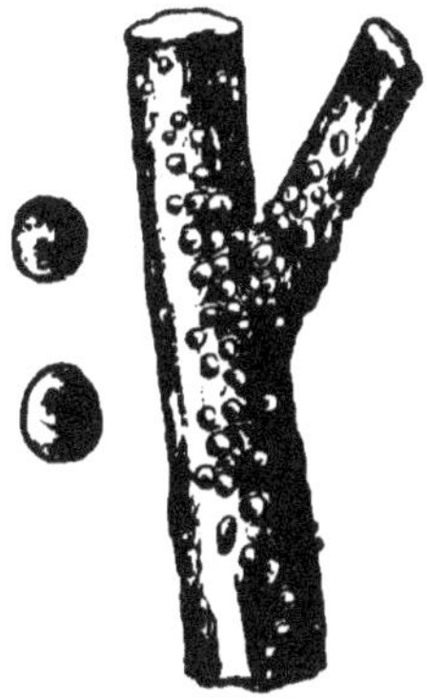

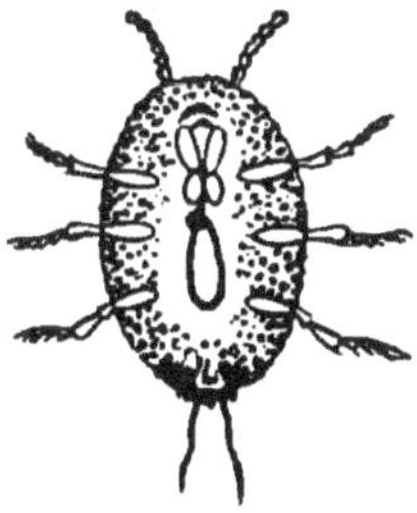

are less than one-hundredth of an inch long; one of them is shown, highly magnified, in Fig. 439.

This scale has been found on olive-trees in various parts of California, but it is said to flourish only on trees in an unhealthy condition, and, as it is chiefly confined to the trunk . and larger limbs, can be easily removed with a stiff brush dipped in a solution of whale-oil soap. It also infests apple and pear trees on the Pacific coast.

423

No. 266.—The Fig-eater.

Allorhina nitida (Linn.).

This beetle, which has acquired the local name of fig-eater in the South, is closely related to the Cetonias, Nos. 81 and 82, which, in the northern portions of the continent, eat the flesh of ripe pears, plums, and peaches. The fig-eater, which is shown in Fig. 440, is a very common insect in the South ;

FIG. 440.

it is nearly an inch long, with a robust body, the wing-cases being velvety green, with light, cream-colored borders. No remedy has been suggested for these insects other than collecting and destroying them.

SYNONYMICAL LIST.

In the following list the older as well as the newer names of the insects referred to are given, as a guide to those who may not have become familiar with the changes which have taken place in insect nomenclature within the past few years. The list does not include all the changes proposed, but only such as have been generally accepted by entomologists, with a few others which have such a weight of testimony in their favor as will probably lead to their general acceptance

1. Schizoneura lanigera (Hausm.).
   Eriosoma pyri Fitch.
   Pemphigus pyri Fitch.
   Aphelinus mali (Hald.).
   Eriophilus mali Hald.
2. Saperda candida Fabr.
   Saperda bivittata Say.
3. Chrysobothris femorata (Fabr.).
   Buprestis femorata Fabr.
4. Leptostylus aculifer (Say).
   Lamia aculifera Say.
6. Monarthrum mali (Fitch).
   Tomicus mali Fitch.
13. Amphicerus bicaudatus (Say).
    Bostrichus bicaudatus Say.
14. Epicærus imbricatus (Say).
    Liparus imbricatus Say.
16. Mytilaspis pomorum Bouché.
    Aspidiotus conchiformis Gmelin.
    Mytilaspis pomicorticis Riley.
    Tyroglyphus malus (Shimer).
    Acarus malus Shimer.
17. Chionaspis furfurus (Fitch).
    Aspidiotus furfurus Fitch.
    Aspidiotus Harrisii Walsh.
20. Pimpla conquisitor (Say).
    Cryptus conquisitor Say.
21. Nemoræa leucaniæ (Kirkp.).
    Exorista leucaniæ Kirkp.

24. Œdemasia concinna (Sm. & Abb.).
    Notodonta concinna Sm. & Abb.
25. Anisopteryx vernata (Peck).
    Phalena vernata Peck.
27. Podisus spinosus (Dallas).
    Arma spinosa Dallas.
28. Platysamia Cecropia (Linn.).
    Attacus Cecropia Linn.
    Smicra mariæ (Riley).
    Chalcis mariæ Riley.
29. Cœlodasys unicornis (Sm. & Abb.).
    Notodonta unicornis Sm. & Abb.
34. Tolype velleda (Stoll).
    Gastropacha velleda Stoll.
35. Cacœcia rosaceana (Harris).
    Lozotænia rosaceana Harris.
36. Teras minuta (Robs.).
    Tortrix malivorana Le Baron.
37. Phycis indigenella (Zeller).
    Acrobasis indigenella Zeller.
    Phycita nebulo Walsh.
    Tachina phycitæ (Le Baron).
    Exorista phycitæ Le Baron.
38. Tmetocera ocellana (Schiff).
    Tortrix ocellana Schiff.
    Penthina oculana Harris.
    Grapholitha oculana Can. Ent.
40. Teras minuta Robs.
    Tortrix Cinderella Riley.

425

41. Phoxopteris nubeculana (Clem.).
   Anchylopera nubeculana Clem.
43. Nolaphana malana (Fitch).
   Brachytænia malana Fitch.
44. Ypsolophus pometellus (Harris).
   Rhinosia pometellus Harris.
   Chœtochilus pometellus Fitch.
45. Agrotis saucia (Hübner).
   Agrotis inermis Harris.
   Agrotis clandestina (Harris).
   Noctua clandestina Harris.
47. Eugonia subsignaria (Hübner).
   Endalinia subsignaria Hübner.
   Ennomos subsignaria Packard.
48. Phobetron pithecium (Sm. & Abb.).
   Limacodes pithecium Sm. & Abb.
55. Odontota rosea (Weber).
   Hispa rosea Weber.
   Hispa marginata Say.
57. Adalia bipunctata (Linn.).
   Coccinella bipunctata Linn.
   Cycloneda sanguinea (Linn.).
   Coccinella sanguinea Linn.
   Coccinella munda Say.
   Megilla maculata (De Geer).
   Coccinella maculata De Geer.
   Hippodamia maculata Muls.
   Anatis 15-punctata (Oliv.).
   Mysia 15-punctata Oliv.
   Harmonia picta (Rand).
   Coccinella picta Rand.
61. Sciara mali (Fitch).
   Molobrus mali Fitch.
64. Lithophane antennata Walker.
   Xylina cinerea Riley.
67. Oncideres cingulatus (Say).
   Saperda cingulata Say.
68. Xyleborus pyri (Peck).
   Scolytus pyri Peck.
   Tomicus pyri Harris.
71. Lygus lineolaris (P. Beauv.).
   Capsus lineolaris P. Beauv.
   Capsus oblineatus Say.
73. Pomphopœa aenea (Say).
   Lytta aenea Say.
77. Cotalpa lanigera (Linn.).
   Areoda lanigera Linn.

81. Euphoria Inda (Linn.).
   Cetonia Inda Linn.
82. Euphoria melancholica (Gory).
   Cetonia melancholica Gory.
84. Apatela occidentalis (G. & R.).
   Acronycta occidentalis G. & R.
85. Apatela superans (Guen.).
   Acronycta superans Guen.
88. Telea polyphemus (Linn.).
   Attacus polyphemus Linn.
95. Coccotorus scutellaris (Lec.).
   Anthonomus prunicida Walsh.
98. Phlœotribus liminaris (Harris).
   Tomicus liminaris Harris.
100. Ithycerus noveboracensis (Forster).
   Ithycerus curculionides Herbst.
101. Ptycholoma persicana (Fitch).
   Crœsia persicana Fitch.
   Lozotænia fragariana Packard.
104. Dicerca divaricata (Say).
   Buprestis divaricata Say.
109. Crepidodera Helxines (Linn.).
   Altica nana Say.
110. Callosamia Promethea (Drury).
   Attacus Promethea Drury.
112. Hyperchiria Io (Linn.).
   Saturnia Io Linn.
114. Cacœcia cerasivorana (Fitch).
   Lozotænia cerasivorana Fitch.
117. Thecla titus Fabr.
   Thecla mopsus Boisd. & Lec.
130. Sinoxylon basilare (Say).
   Apate basilaris Say.
131. Ampeloglypter Sesostris (Lec.).
   Baridius Sesostris Lec.
   Madarus vitis Riley.
132. Darapsa myron (Cramer).
   Chœrocampa pampinatrix Sm.
133. Philampelus Pandorus (Hübner).
   Philampelus satellitia Linn.
144. Oxyptilus periscelidactylus
      (Fitch).
   Pterophorus periscelidactylus
      Fitch.
147. Pyrophila pyramidoides (Guen.).
   Amphipyra pyramidoides Guen.
148. Pyrophila tragopoginis (Linn.).
   Agrotis repressus Grote.

150. Graptodera chalybea (Illig.).
   Haltica chalybea Illig.
152. Fidia longipes (Mels.)
   Pachnephorus longipes Mels.
157. Erythroneura vitis (Harris).
   Tettigonia vitis Harris.
165. Cyrtophyllus concavus (Harris).
   Platyphyllum concavum Harris.
   Phylloptera oblongifolia (De Geer).
   Locusta oblongifolia De Geer.
171. Eudemis botrana (Schiff).
   Penthina vitivorana (W. & R.).
172. Craponius inæqualis (Say).
   Ceutorhynchus inæqualis Say.
174. Bembecia marginata Harris.
   Ægeria rubi Riley.
176. Oberea bimaculata Oliv.
   Oberea tripunctata Fabr.
181. Apatela brumosa Guen.
   Acronycta verrillii Grote.
183. Chelymorpha Argus Leich.
   Chelymorpha cribraria Fabr.
184. Synchlora rubivoraria (Riley).
   Aplodes rubivora Riley.
191. Tyloderma fragariæ (Riley).
   Analcis fragariæ Riley.
192. Phoxopteris comptana Frol.
   Anchylopera fragariæ W. & R.
193. Eccopsis permundana (Clemens).
   Exartema permundana Clemens.
194. Apatela oblinita (Sm. & Abb.).
   Acronycta oblinita Sm. & Abb.
195. Agrotis Ypsilon (Rott.).
   Agrotis suffusa D. & S.
   Agrotis telifera Harris.
   Agrotis tricosa Lintner.
   Agrotis jaculifera Guen.
   Hadena devastatrix (Brace).
   Agrotis devastator Harris.
196. Paria sex-notata (Say).
   Colaspis sex-notata Say.
197. Phyllotreta vittata (Fabr.).
   Crioceris vittata Fabr.

   Phyllotreta striolata Illig.
   Haltica striolata Harris.
203. Psenocerus supernotatus (Say).
   Clytus supernotatus Say.
208. Eufitchia ribearia (Fitch).
   Ellopia ribearia Fitch.
209. Grapta progne (Cram.).
   Vanessa progne Cram.
212. Pœcilocapsus lineatus (Fabr.).
   Lygæus lineatus Fabr.
   Capsus 4-vittatus Say.
215. Epochra Canadensis (Loew).
   Trypeta Canadensis Loew.
216. Endropia armataria (Herr. Sch.).
   Priocycla armataria Herr. Sch.
219. Dakruma convolutella (Hübn.).
   Zophodia convolutella Hübn.
   Pempelia grossulariæ Packard.
   Myelois convolutella Packard.
223. Crepidodera cucumeris (Harris).
   Haltica cucumeris Harris.
224. Eudioptis hyalinata (Linn.).
   Phakellura hyalinatalis Linn.
225. Eudioptis nitidalis (Cram.).
   Phakellura nitidalis Cram.
227. Rhopobota vacciniana (Packard).
   Anchylopera vacciniana
   Packard.
228. Teras oxycoccana (Packard).
   Tortrix oxycoccana Packard.
229. Teras vacciniivorana (Packard).
   Tortrix vacciniivorana Packard.
239. Papilio cresphontes Fabr.
   Papilio thoas Boisd.
241. Platynota rostrana (Walker).
   Teras rostrana Walker.
258. Ceroplastes Floridensis Comstock.
   Ceroplastes rusci Linn.
   (Ashmead)
264. Cycloneda abdominalis (Say).
   Coccinella abdominalis Say.
266. Allorhina nitida (Linn.).
   Cotinis nitida Linn.

INDEX.

429

THE END.